ENCYCLOPAEDIA OF MOLECULAR BIOLOGY

ENCYCLOPAEDIA OF MOLECULAR BIOLOGY

(Vol. 3)

A. Tyagi

ANMOL PUBLICATIONS PVT. LTD.

NEW DELHI - 110 002 (INDIA)

ANMOL PUBLICATIONS PVT. LTD.
H.O.: 4374/4B, Ansari Road, Darya Ganj,
New Delhi-110 002 (India)
Ph.: 23261597, 23278000

B.O.: No. 1015, Ist Main Road, BSK IIIrd Stage
IIIrd Phase, IIIrd Block,
Bangalore - 560 085 (India)
Visit us at: www.anmolpublications.com

Encyclopaedia of Molecular Biology

First Published, 2006

ISBN 81-261-2677-9 (Set)

[Responsibility for the facts stated, opinions expressed, conclusions reached and plagiarism, if any, in these volumes is entirely that of the Author. The Publishers bear no responsibility for them, whatsoever.]

PRINTED IN INDIA

Published by J.L. Kumar for Anmol Publications Pvt. Ltd., New Delhi - 110 002 and Printed at Mehra Offset Press, Delhi.

Contents

Preface

Smallest fundamental unit of an element or a compound is called a molecule. The rapid progress and multi-faceted development of Science, in general; and of Biology, in particular, has ushered in an absolutely new era in the history of mankind, regarding the physio-chemical department of life.

There has been an urge to know more and more from the day one. It is due to this urge that man has been scaling ever new heights day-after-day, be it related to Science or Commerce or Politics and so on. Is this whole endeavour, which has proved to be a real boon for man, or in other words, the greatest gift, the twentieth century gave to humanity ? It really brought about a revolution in the whole human life. The emergence of Molecular Biology, as a branch of Science for study, could also be possible by virtue of this very present of the past.

The discipline is very vast and with the passage of time, big breakthroughs are in the offing. Feeling the pulse of the time, it deemed suitable, rather, necessary to bring out a comprehensive and compact encyclopaedia on this subject in order to cater to the maximum requirements of the lovers of Molecular Biology.

Comprising five volumes, this trivial attempt is expected to prove to be an asset for Science elite and students alike. Now, savour it molecule-by-molecule.

Editor

1

The Macromolecules

Cells are actually regarded as macromolecular assemblies. These macromolecules, making 30% of the cell's mass (70% of cell's mass is water), include proteins, nucleic acids and polysaccharides. Though lipids are also large molecules, they are not large enough to be classified as macromolecules. Among the classes of macromolecules, proteins and nucleic acids are very large molecules, responsible for conveying genetic information. While nucleic acids carry the genetic information from one cell generation to another, proteins provide a means of expressing this genetic information. Polysaccharides are also big enough to be classified as macromolecules, but they do not have the complexity of proteins and nucleic acids to carry or express the information. Of the three classes of macromolecules, proteins have the highest of diversity, nucleic acids have some diversity and polysaccharides have the lowest level of diversity. These macromolecules are polymers of small micromolecules described in the previous chapter (proteins are polymers of amino acids, nucleic acids are polymers of nucleotides and polysaccharides are polymers of monosaccharide sugar molecules).

Each macromolecule has (i) a backbone consisting of a regularly repeating series of bonds and (ii) side groups branching out from the backbone and providing characteristic diversity. The bonds connecting the monomers and the side groups known in the above three macromolecules are presented in Table and Figure. In this chapter, we discuss some details of the structure of these macromolecules. Their synthesis will be described in chapters elsewhere in the book.

Bonds in the Backbone and the Side Groups of Biological Macromolecules.

Macromolecule	*Nature of backbone*	*Nature of side groups*
1. Proteins	Peptide	Amino acids (20)
2. Nucleic acids	Sugar-phosphate bond (phosphodiester bond)	Nitrogenous bases (4)
3. Polysaccharides	Glycosidic bond	Hydroxyl group or substituent (–)

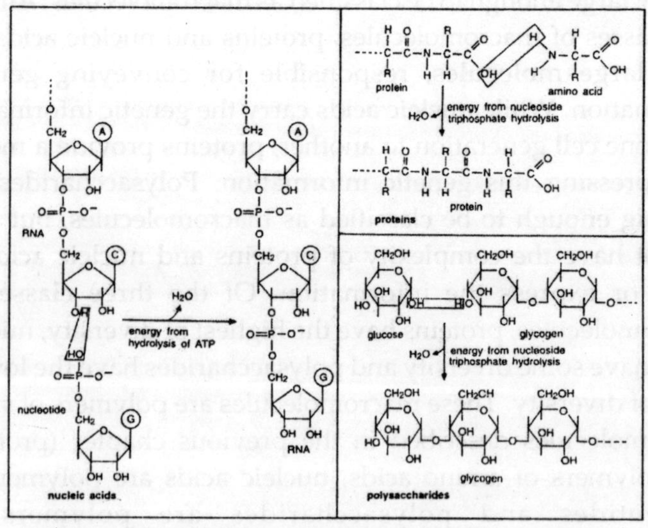

Derivation of three types of macromolecules(polymers)from their monomers due to formation of three kinds of bonds.

Covalent and Weak Non-covalent Bonds in Macro-Molecules : The structure of macromolecules is achieved and maintained with the help of strong covalent bonds and some weak non-covalent bonds. Therefore it is necessary to understand the nature of these bonds, before we discuss the structure of macromolecules. The monomers in a macromolecular chain (a polymer) are held together by covalent bonds, which are formed between atoms due to sharing of one or more electrons (in a single bond, one electron from each atom is shared; in a double bond a total of four electrons are shared). Atoms joined by two or more covalent bonds can not rotate freely about the bond axis and this influences three-dimensional shape of many macromolecules. These covalent bonds are strong enough (particularly in water) to preserve the sequence of subunits for a long time (Table); their formation may need enzymes, and their strength is not affected by the environment.

The weak bonds, on the other hand, are formed between different parts of the same macromolecule or between different macromolecules. Their formation, which does not need any enzymes, is strongly influenced by the aqueous environment. Thus these weak bonds determine not only the secondary and tertiary structures of these macromolecules, but also determine

Covalent Bonds and Non-Covalent Bonds Differ in Strength.

Type of bond	*Length of bond (nm)*	*Strength of bond (kcal/mole) (energy required tobreak a bond)*	
		in vacuum	*in water*
1. Covalent	0.15	90	90
2. Ionic	0.25	80	3
3. Hydrogen	0.30	4	1
4. Van der Waals (per-atom)	0.35	1.0	0.1
5. Hydrophobic	—	3	0.5

their distribution and arrangement within a cell. These weak bonds thus are indispensable for the very existence of the cell. In view of this, an understanding of these weak interactions is necessary to understand the structure and function of macromolecules.

Spontaneous formation of a bond always involves release of energy, and breaking of this bond will similarly consume energy; the stronger the bond, greater the amount of energy released. In a cell, bonds are regularly formed and broken, so that at an equilibrium the number of bonds formed is equal to the number of bonds broken. The proportion of bonded molecules is described by the following formula, where Keq is the equilibrium constant.

$$A + B \longrightarrow AB + \text{energy}$$

$$K_{eq} = \frac{\text{Conc. of AB}}{\text{Conc. of A} \times \text{Conc. of B}}$$

At this equilibrium, the change in free energy is zero $(\Delta G = 0)$ and the amount of this free energy is minimum. (Free energy is the energy that has the ability to do work).

Four types of non-covalent bonds that occur in macromolecules include (i) ionic bonds, (ii) hydrogen bonds, (iii) Van der Waals forces, (iv) hydrophobic interactions.

Ionic Interactions : Ionic interactions occur between oppositely charged groups (ions). These interactions may occur between fully charged groups (ionic bonds) or between partially charged groups. The force of attraction between the charges q^+ and q^- is defined by Coulomb's law, which says that force $= (q^+q-)/r^2$ D (where D = dielectric constant, which is 1 for vacuum and 80 for water; r = distance of separation). In the absence of water, in solids these forces are very strong as in salt (NaCl) or marble.

In NaCl, the force between Na^+ and Cl^- has a strength comparable to covalent bond. However in solution, water molecules interact with charged group in such a way that water shields the charge, thus weakening the ionic bond. In a protein, acidic and basic amino acids may also interact forming ionic bonds; similarly a protein molecule having a charged group may interact with another protein molecule (protein-protein interaction) or other target molecule like nucleic acid (protein-nucleic acid interaction) having an opposite charge. Another example of ionic interactions is that between an enzyme and its substrate.

Hydrogen Bonds : Hydrogen bonds are weak electrostatic bonds, in which a hydrogen atom is shared between two other atoms, both electronegative, as between O and N or between O and O in the following examples.

> C = O — — — — — — — — H – N–
hydrogen bond convalent bond

(~0.2 nm) (~0.1 nm)

> C = O — — — — — — — — H – O–

In proteins, hydrogens bonds are formed between NH and CO groups of the peptide backbone. Hydrogen bonds may also involve side groups in a protein chain. The commonest example of hydrogen bonds is the bonds between different molecules of liquid water (Fig.) or between water molecules and other molecules. Hydrogen bonds also occur between amino acids in a polypeptide chain or between hydroxyl groups of two different molecules (Fig.).

Many organic molecules also possess ionic groups with charges which are neutralized by nearby, oppositely charged groups within these organic molecules or by inorganic ions (Na^+ , K^+, Mg^{++}, Cl^-, SO_4^-). But, directional bonds are formed, if oppositely charged groups can form hydrogen bonds (e.g. COO^- and NH^+_3 may be held together by hydrogen bonds;

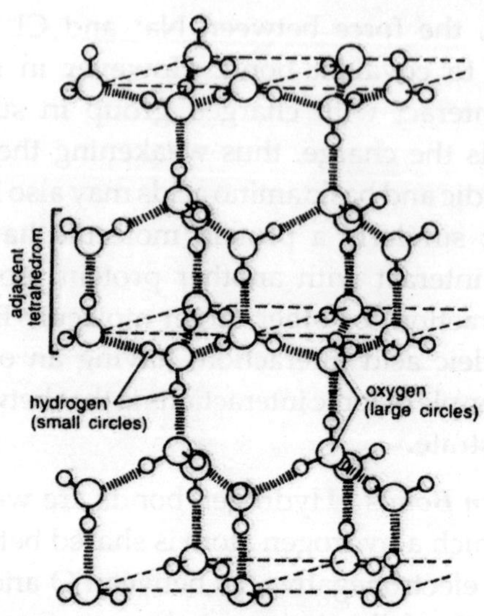

Formation of hydrogen bond between different molecules of water.

when such hydrogen bonds are formed between ionic groups, ionic bonds are hydrogen bonds; Fig.).

Van der Waals Bonding : Van der Waals bonding results due to nonspecific attractive force between atoms which come close to each other. In this case there are no permanent charges associated with participating groups, but there are fluctuating charges induced due to nearness of molecules.

Therefore these bonds occur among nonpolar as well as polar molecules and their strength depends on the distance between interacting groups (Fig.). If these atoms are brought very close to each other, they will very strongly repel each other and this Van der Waals repulsion plays a major role in maintaining the conformation of biological molecules. However, at a particular distance between atoms (which is specific for each type of atom), attraction and repulsion forces balance each other (Fig.). This distance is called Van der Waals radius (Table). Interaction between antigen and antibody may also be due to such Van der Waals forces.

(a) hydrogen bond between peptide groups

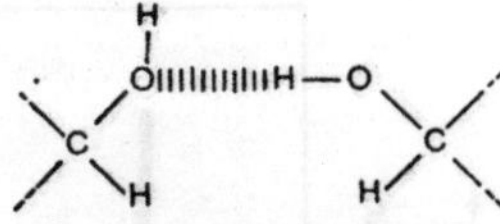

(b) hydrogen bond between two hydroxyl groups

(c) hydrogen bond between a charged carboxyl group and the hydroxyl group of tyrosine

(d) hydrogen bond between a charged amino group and a charged carboxyl group

Formation of hydrogen bonds (a) between amino acids in a polypeptide, (b) between two hydroxyl groups, (c) between a carboxyl group and a hydroxyl group and (d) between amino group and a carboxyl group

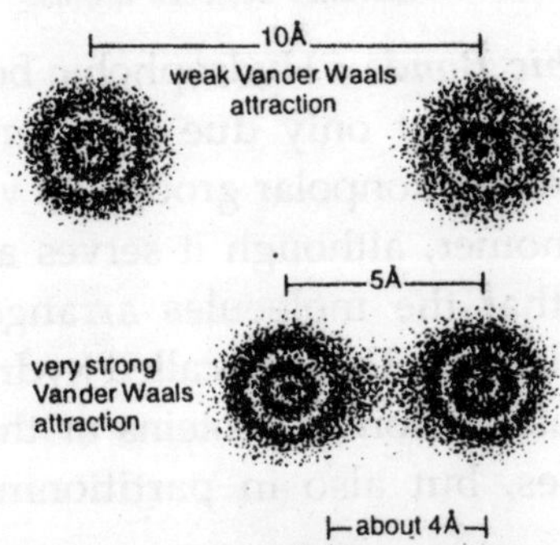

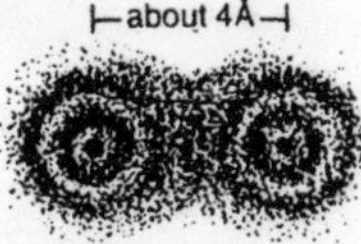

Van der Waals bonds, showing dependence of their strength on the distance between interacting groups.

Van der Waals radii of atoms in biological molecules.

Atom	*Van der Walf's radius*	*Atom*	*Van der Wall's radius*
H	1.2 Å	P	1.90 Å
N	1.5 Å	S	1.85 Å
O	1.4 Å	C	2.00 Å

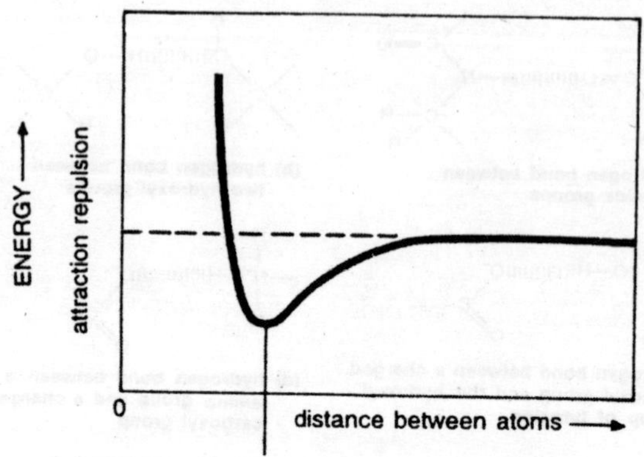

Relationship between Van der Waals force (attraction/repulsion) and the distance between atoms.

Hydrophobic Bonds : Hydrophobic bonds result not due to any attraction, but only due to a tendency of water to exclude or repel the nonpolar groups. In view of this the term bond is a misnomer, although it serves a useful purpose by emphasizing that the molecules arrange to avoid contact with water molecules. These so called hydrophobic bonds help not only in stabilization of proteins or their complexes with other molecules, but also in partitioning of proteins into membranes.

Nucleic Acids

Nucleic acids are the most important macromolecules in the cells of all organisms. In the beginning of the 20th century it was shown that chromosomes (consisting mainly of nucleic acids and proteins) within a cell carry the hereditary factors. However, for several decades, we did not know whether nucleic acids or proteins are the carriers of genetic information. In 1940's and 1950's, several experiments demonstrated that nucleic acids are the genetic material within the cells. Some of these experiments will be briefly discussed before we discuss the structure of these macromolecules.

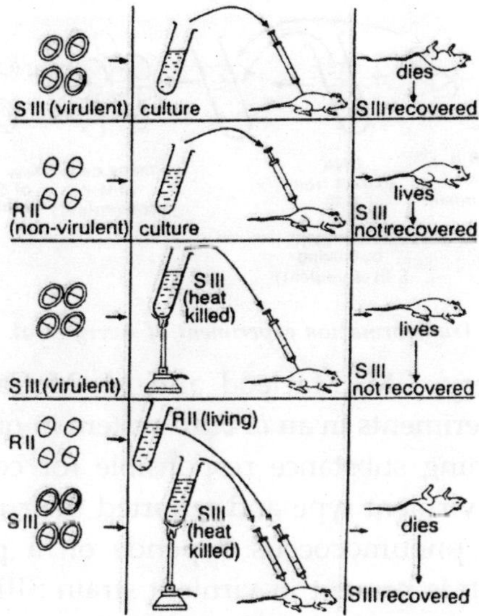

Griffith's experiment demosntrating transformation in pneumococcus.

Nucleic Acids as Genetic Material : Chromosomes mainly consist of nucleoproteins having two components, nucleic acids and proteins. One of these should obviously constitute the genetic material. Experiments demonstrating that nucleic acids and not the proteins contain the genetic information will be briefly discussed in this section.

Transformation Experiments : Transformation experiments were initially conducted by F. Griffith in 1928. He injected a mixture of two strains of pneumococcus *(Diplococcus pneumoniae)* into mice. One of these two strains, SIII was virulent and the other strain RII was non virulent (causing no infection). Heat killed virulent strain SIII, when injected individually did not cause death, showing that infectivity after heat killing is lost.

The mice injected with a mixture of RII (living) and SIII (heat killed) died and virulent pneumococci could be isolated from these mice (Fig.). This phenomenon was described as transformation.

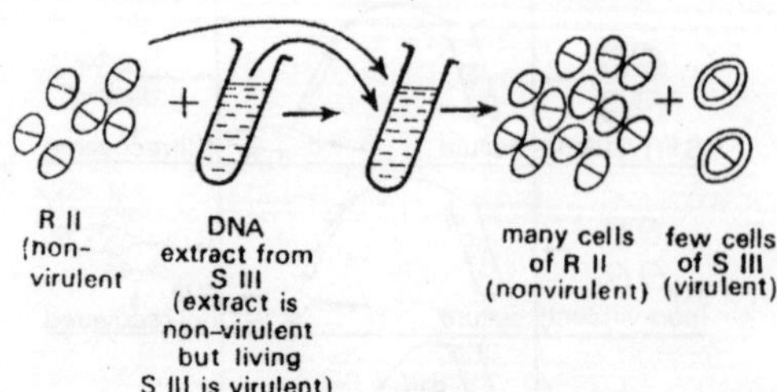

Transformation experiment of Avery et al, (1944).

O.T. Avery, CM. Macleod and M. McCarthy repeated Griffith's experiments in an *in vitro* system in order to identify the transforming substance responsible for converting non-virulent into virulent type and reported their results in 1944. Virulence in pneumococcus depends on a polysaccharide capsule which is present in virulent strain SIII and is absent in non- virulent strain RII. The cells of non-capsulated type (RII) were treated with an extract of DNA from capsulated strain SIII. A few cells of SIII type could be isolated from the mixture (Fig.). This phenomenon of transferring characters of one strain to another by using a DNA extract of the former is called transformation. When the extract was treated with DNAase (an enzyme which destroys DNA) this transforming ability was lost. Protesases (enzymes which destroy proteins) did not affect the transforming ability. These experiments thus indicated that DNA and not the proteins is the genetic material.

Experiments with Bacteriophage (T2) Infection : The fact that DNA is the genetic material could also be demonstrated with the help of infecting bacteria by bacteriophage known as T2. This phage consists of proteins and DNA. The complex structure of a T2 phage particle is shown is Figures. It consists of a head and a tail. The head consists of protein coat and DNA. The tail similarly consists of tail core, tail sheath and tail fibres. The phage attaches to the host bacterium with the help of tail fibres. After infection, multiplication of phage occurs and a large number of progeny phages, genetically similar to the infecting phage are produced.

In 1952, A.D. Hershey and M.J. Chase reported results of experiments which demonstrated that only DNA of the phage enters host cells and that this DNA carries all the genetic

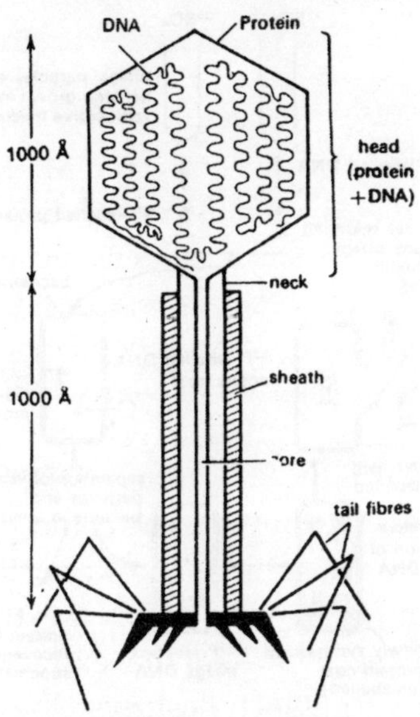

Structure of T2 bacteriophage.

information necessary for assembly of new phage progeny. In these experiments, phage DNA was made radioactive by growing infected bacteria on a medium containing $^{32}PO_4$. Since phage proteins do not contain phosphorus, only DNA would be labelled. Similarly, phage proteins were labelled with the help of $^{35}SO_4$. Since DNA does not contain sulphur, only protein would be labelled with ^{35}S. Such differential labelling would enable one to distinguish between DNA and proteins of the phage without performing any chemical tests. Hershey and Chase then allowed both kinds of labelled phage particles to infect bacteria. The infected bacteria were immediately agitated in a waring blender. After shaking, only radioactive ^{32}P was found associated with bacterial cells and ^{35}S was found only in surrounding medium and not in bacterial cells. When phage progeny was studied for radioactivity in this experiment, it was found that the phage progeny carried label only with ^{32}P. The progeny was not labelled with ^{35}S. This experiment is

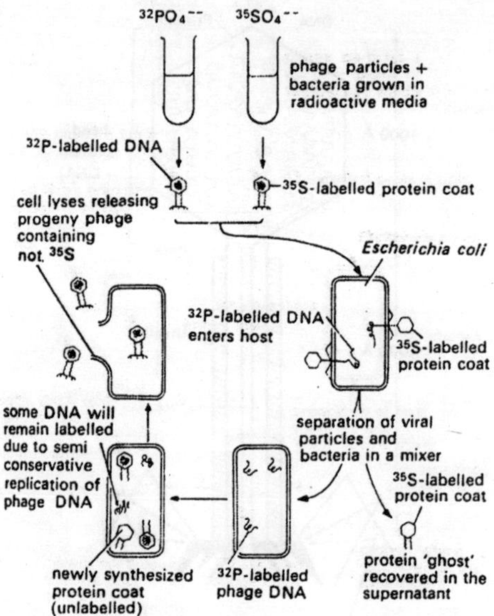

Experiment of Hershey and Chase, where DNA of bacteriophage is labelled with 32p and the protein is labelled with ^{35}S.

diagrammatically represented in Figure and clearly indicated that only DNA and not protein enters the bacterial cells. The protein coat is left outside. This DNA entering the host cell carries all the genetic information, for synthesis of new phage particles. Indirectly, this experiment conducted by Hershey and Chase proved that DNA is the sole genetic material in DNA bacteriophages. A.D. Hershey shared with M. Dellbruck and S.E. Luria, the Nobel Prize in Medicine/Physiology for 1969.

Experiments with tobacco mosaic virus (TMV). With the help of transformation experiment and bacteriophage infection discussed above, it was shown that DNA is the genetic material. The first evidence that RNA also has the capacity to carry genetic information came from experiments conducted with tobacco mosaic virus (TMV). This virus does not contain any DNA, and is composed of RNA (6%) surrounded by a hollow cylinder of protein subunits. The experiments conducted by H. Fraenkel-Conrat in California showing that RNA is the genetic material in this virus will be briefly discussed in this section.

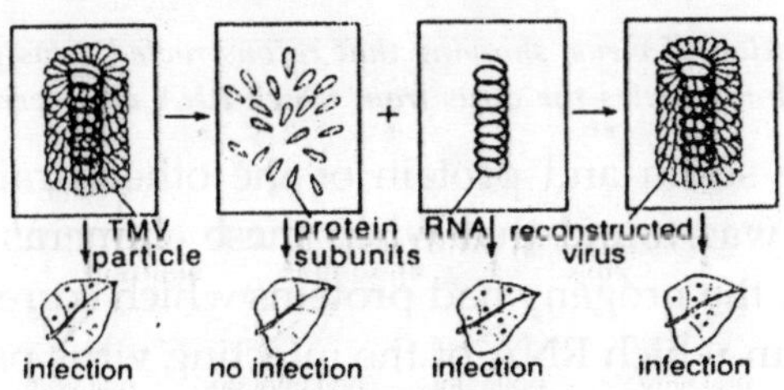

Experiment of Fraenkel-Conrat on TMV showing that RNA and not the protein can cause infection.

Techniques were first developed for separating TMV particles into RNA and proteins. Later by using RNA and proteins separately in tests for infectivity, it could be shown that RNA alone was able to cause infection. Such a property was not found in the protein fraction (Fig.). It was also possible to synthesize chimeric virus particles using RNA from one strain and protein from another strain. Such chimeras had serological properties of those strains from which protein was derived, but in other properties, it resembled the other strain from which RNA was used.

In one experiment, two viruses used were tobacco mosaic virus (TMV) and Holmes rib-grass virus (HRV). The latter was isolated from *Plantago lanceolata* and therefore, is also known as *Plantago* strain. Proteins of these two viruses differ in having different frequencies and sequences of amino acids. Also these viruses give different symptoms. On leaves of a particular Variety of tobacco, while TMV produces mottling of leaves, HRV produces distinct ring patterns. Reciprocal chimeras using

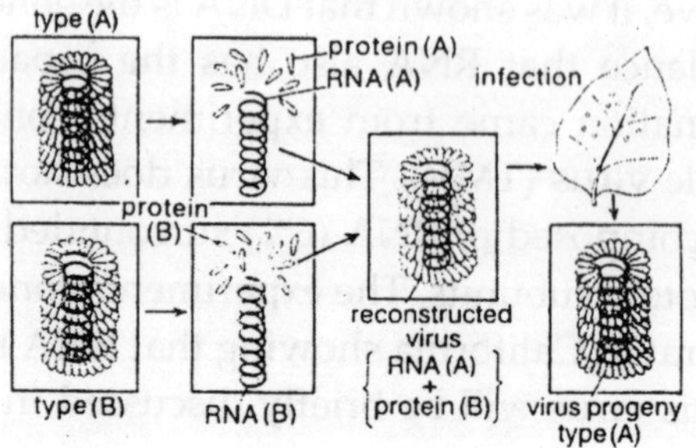

Reconstruction of virus, showing that reconstructed virus gives progeny, which resembles the virus from which RNA was derived.

RNA of one strain and protein of the other strain could be obtained. It was found that when these chimeras were used for infection, the progeny had proteins which corresponded to the virus from which RNA of the infecting virus particles was derived. This proved that specificity of virus proteins was determined by RNA alone and the proteins carried no information. Results of these experiments are diagrammatically represented in Figure.

Subsequently, after the demonstration of genetic nature of RNA in TMV, RNA was found in a number of bacteriophages. Therefore, it may be concluded that nucleic acids in general, RNA as well as DNA, carry genetic information. However, DNA perhaps always functions as genetic material ; but RNA in majority of cases is non-genetic; only in specific systems where DNA is absent, it functions as genetic material. In recent years, RNA has been shown to functions as enzymes (ribozymes) also. This aspect of RNA led to award of Nobel Prize in chemistry in 1989 (T. Cech and S. Altman). 'Prions' have also been described in recent years as 'proteinaceous infectious particles' which is an example of 'protein conformation based inheritance'. Nobel Prize for 1997 was awarded to Stanley B. Prusiner for the discovery of 'prions'.

Structure of Nucleic Acids (Polynucleotides) : Nucleic acid molecule is a long chain polymer (polynucleotide) composed of monomeric units, called nucleotides, described in the previous chapter. These nucleotide monomer units are joined by the formation of phosphodiester bonds (a diester bond is one which involves two ester bonds). A phosphodiester bond will be formed between any two adjacent nucleotides.

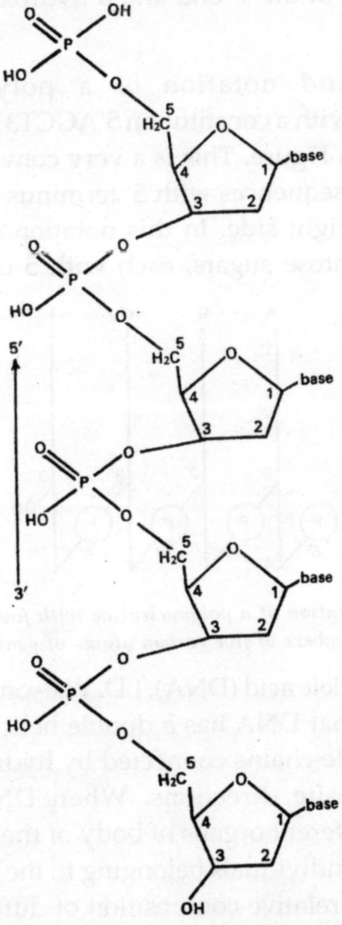

A polynucleotide chain showing phosphodiester bonds.

A polynucleotide chain (Fig.) would have a direction so that, if it starts from C^3 it would end in C^5 and if it starts from C^5 it would end in C^3. Normally, therefore, whether we are referring to a mononucleotide or else to a polynucleotide we should indicate the groups present at C^3 end and at C^5 end. It could be either a phosphate group or a hydroxyl group. For instance, 5′pApT.OH3' dinucleotide would mean that it is a polynucleotide with only two nucleotide units and has phosphate group at the 5^0 end and a hydroxyl group at the 3' end.

A short hand notation of a polynucleotide (or oligonucleotide) with a constitution 5′AGCT3' or 5′pApGpCpT-OH3' is shown in Figure. This is a very convenient method to write nucleotide sequences with 5' terminus on the left and 3' terminus or the right side. In this notation the Vertical lines represent the pentose sugars, each with 5 carbon atoms.

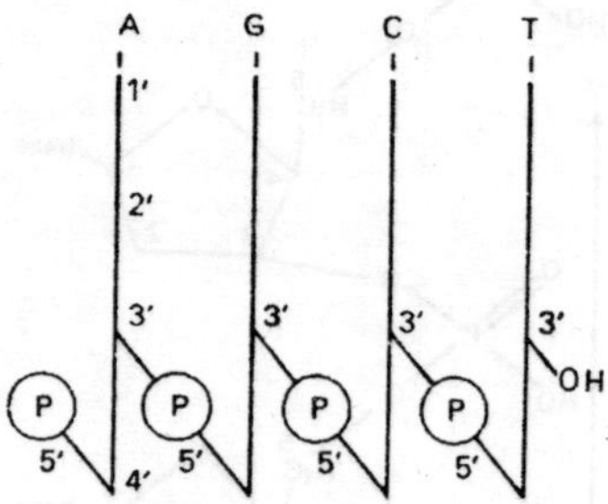

A short-hand notation of a polynucleotide with four nucleotides (1′to 5′ represent numbers of five carbon atoms of pentose sugar).

Deoxyribonucleic acid (DNA). J.D. Watson and F.H.C. Crick showed in 1953 that DNA has a double helical structure with two polynucleotide chains connected by hydrogen bonds and running in opposite directions. When DNA samples are analysed from different organs of body of the same individual or from different individuals belonging to the same species, no differences in the relative composition of different bases were observed. The following generalizations could also be made : (i) Number of adenine molecules always equalled number of thymine molecules, (ii) Number of cytosine molecules always equalled number of guanine molecules, (iii) Number of adenine molecules did not necessarily equal that of guanine and similar relationship was observed between cytosine and thymine.

Although A = T and C = G, there is no restriction on the ratio (A + T)/(C + G); nor is there any restriction on sequence of bases in one polynucleotide chain. But, since A is always linked to T and C to G as determined from the above evidences, sequence of bases in one polynucleotide should determine the sequence of bases in the other polynucleotide of the double helix.

Besides chemical analysis, DNA was also studied by X-ray crystallography by M.H.F. Wilkins and his associates. Such studies demonstrated that DNA was a helical structure with a diameter of 20 A° and a pitch (one round) of about 34 A°. On the basis of these studies of chemical analysis and x-ray

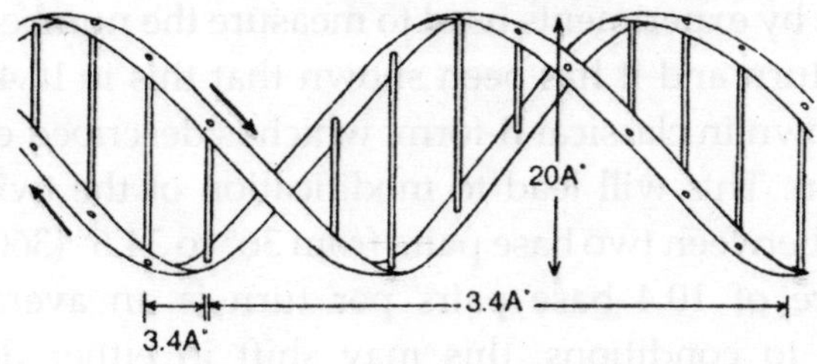

A double helix model of DNA as proposed by Watson and Crick.

crystallography, following features of the DNA molecule are now known: (i) DNA consists of a double helix, in which two polynucleotide chains are coiled about the same axis in such a fashion that they can separate from one another only by uncoiling and that lateral separation would not be possible, (ii) Bases are set in a plane at right angle to the long axis, (iii) Two polynucleotide chains which run in opposite directions, have complementary base sequences, since adenine is linked to thymine and cytosine to guanine. (iv) Distance between two base pairs is 3.4 A° and there are 10 base pairs in each turn.

Choice of base pairing is understandable because of a constant diameter of 20 A° in the molecule; two purines would need too much space and two pyrimidines would occupy too little space to pair. When a pyrimidine always pairs with a purine, space could remain constant. The pattern of coiling of two polynucleotide chains is shown in Figure.

Alternative forms of DNA Double Helices : The well known structure of DNA described above pertains to DNA structure, described by Watson and Crick (1953). Helical coiling of this DNA is assumed to be right handed and this form of DNA has been called B-form. It has been shown that DNA may also exist in other alternative forms of double helical structures. These alternative forms differ in features like (i) number of residues per turn ('n') or (ii) the spacing of residues along the helical axis ('h'). The existence of DNA as a double helix has also been confirmed by experiments used to measure the number of base pairs per turn and it has been shown that this is 10.4 instead of 10.0 shown in classical B-form, which is described earlier in this section. This will lead to modification of the twist angle (rotation) between two base pairs from 36° to 34.6° (360°/10.4). This figure of 10.4 base pairs per turn is an average and according to conditions, this may shift in either direction.

Several Forms of DNA Double Helical Structures.

Helix type	*Conditions*	*Base pairs per turn*	*Rotation per bp*	*Vertical rise per bp (h)*	*Helical diameter*
A	75% rel. humidity; Na^+ K^+, Cs^+ ions	11	+32.7° (right handed)	2.3 A	25.5 A
B	92% rel. humidity; low ioinc strength	10	+36.0° (right handed)	3.4 A	23.7 A
C	66% rel. humidity; Li^+ ions	9.33	+ 38.6° (right handed)	3.3 A	23.7 A
Z	Very high salt concentration	12	-30.0° (left handed)	5.7 A	18.4 A

It is thus obvious that there may be several families of structures of DNA, each of a characteristic type showing variation in the values of 'n' and 'h' described above. Three structural forms, namely A, B and C are known for a long time and transitions between them can occur, General characteristics of these three forms are given in Table.

The B-form is found at a very high (92%) relative humidity and low ionic strength. Similarly, A-form is found at 75% relative humidity in the presence of Na^+ K^+, or Cs^+ ions and C-form is found at 60% relative humidity in the presence of Li^+ (Lithium) ions. These three forms are assumed to be found in all DNAs.

There are still other forms called D-form and E-form found rarely as extreme variants with only 8 and 7 ½ base pairs per turn respectively. These are found only in some DNA molecules lacking guanine. In contrast A, B, and C are found in all DNA molecules irrespective of DNA sequence. As shown in Table, B DNA is not the only right handed DNA. Similarly, it has been shown that B DNA is not always right handed. B DNA can also be left handed.

Z-DNA, a Left Handed DNA form : Another form of DNA has been obtained artificially by synthesizing $d(C\text{-}G)_3$ molecules which were isolated as crystals. This form of DNA has also been detected in living cells using specific antibodies raised against it. Because this form of DNA follows a zig-zag course, it has been called Z-DNA. It resembles the B form in many respects and differs from it in many other respects (Table). The differences and similarities are described below:

Resemblances between z-DNA and B-DNA

(i) Both are double helical,

(ii) Two strands of double helix are antiparallel in both DNAs.

(iii) Both forms exhibit A=T and G≡C pairing.

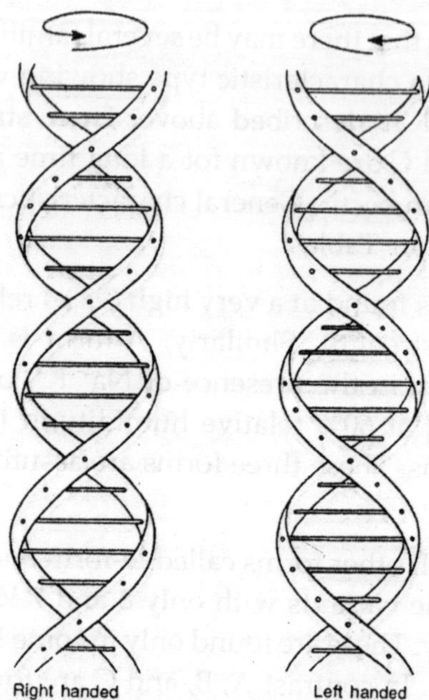

Two helices, one showing right handed sense as found in B-DNA and the other showing left handed sense as found in Z- DNA.

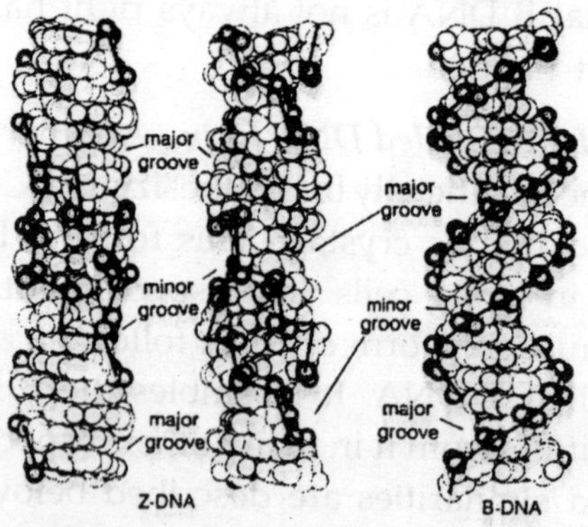

Side views of Z-DNA and B-DNA. Two views of Z-DNA are 30° apart. Irregularity of backbone in Z-DNA is shown by heavy lines showing the path of phosphate residues, which is quite regular and smooth in B-DNA (redrawn from Nature, Vol. 282, 1979).

Differences between z-DNA and B-DNA : (i) Z-DNA has left-handed helical sense as against right handed helical sense of B-DNA (Fig.). (ii) Due to a different arrangement of molecules within Z-DNA polymer, phosphate backnone follows a zig-zag course, while in B- DNA it is regular (Fig.). (iii) In Z-DNA, sugar residues of a polynucleotide chain have alternating orientation (Fig.) so that repeating unit is a dinucleotide as against B- DNA where repeating unit is a monoucleotide and the orientation of sugar molecules is not alternating, (iv) In Z-DNA, one complete helix i.e. a twist through 360°, has twelve base pairs or six repeating dinucleotide units (12 base pairs), while in B-DNA, one complete helix has only ten base pairs or ten repeating units, (v) Because twelve base pairs are accommodated in one helix in Z-DNA, as against ten in B-DNA, the angle of twist per repeating unit (dinucleotide) is 60° as against 36° in B-DNA. (vi) One complete helix is 45 A° in Z-DNA, while it is 34 A° in B-DNA. (vii) Since bases get more length to spread out in Z-DNA and since the angle of tilt is 60°, they are closer to the axis and hence the diameter of Z-DNA molecule is 18 A°, whereas it is 20 A° in B-DNA.

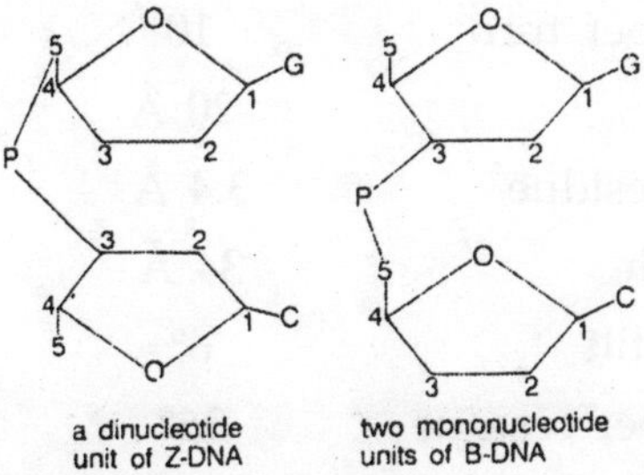

a dinucleotide unit of Z-DNA

two mononucleotide units of B-DNA

Orientation of adjacent sugar residues in Z-DNA and B- DNA showing opposite orientations in Z-DNA and same orientation in B-DNA. This results in dinucleotide units in Z-DNA as against mononucleotide units in B-DNA.

In addition to the above, other differences include (i) differences in glycosidic torsion angle, which is *anti* in B-DNA and *syn* in Z-DNA for deoxyguanosine and (ii) differences in sugar pucker which is C_2' endo for B-DNA and C_3' endo for Z-DNA in deoxyguanosine. The situation is reverse for deoxycytidine. These differences are summarized in Table .

In Vivo Existence and Role of Z-DNA ? A few years ago, it was suggested that Z-DNA may be transient configuration of -DNA in living cells and may thus play a role in (i) regulation of gene activity and/or in (ii) genetic recombination. Earlier the role of Z-DNA in regulation of gene activity was suggested, since Z-DNA specific antibodies could bind specifically to interband regions of salivary gland chromosomes and not to active banded regions. This suggested that Z-DNA conformation renders the DNA inactive. These observations were later shown to be an artefact of chromosome fixation. However, sequences capable of forming Z-DNA are believed to play a role in recombination or rearrangement of DNA (e.g. immunoglobulin gênes).

Comparison of Helices of B-DNA and Z-DNA (from Nature, 282, 1979).

	B-DNA	*Z-DNA*
1. Helix sense	Right handed	Left handed
2. Residues per turn	10	12 (6 dimers)
3. Diameter	-20 Å	-18 Å
4. Rise per residue	3.4 Å	3.75 Å
5. Helix pitch	34 Å	45 Å
6. Base pair tilt	6°	7°
7. Rotation per residue	36°	60° (per dimer)
8. Glycoside torsion angle,		
deoxyguanosine	Anti	Syn
deoxycytidine	Anti	Anti
9. Sugar pucker,		
deoxyguanosie	C_2'endo	C_3'endo
deoxycytidine	C_3' endo	C_2'endo
10. Distance of P from axis,		
dGpC	9.0 Å	8.0 Å
dCpG	9.0 Å	6.9 Å

RL Model by Sasisekharan : During 1976-80, V. Sasisekharan and his co-workers, of Indian Institute of Science, Bangalore, emphasized that the right handed form (R) of double helical B-DNA can not easily explain the unwinding of the duplex due to topological difficulties. They suggested that right handed (R) and left handed (L) double helices are equally likely and therefore proposed a new R1 model. In this model a B-DNA structure with alternating right handed and left handed segments of approximately five base pairs, in a repeat of ten base pairs, was proposed. This model provides for conformational flexibility. The essential differences between the right handed (R) double helix of Watson and Crick and the R1 model of Sasisekharan are given in Table. There is strong experimental evidence available in favour of R1 model, although one does not know the possible maximum continuous lengths of right handed and left handed segments of DNA.

Supercoils in Closed DNA : The double helix structure of DNA represents a linear DNA molecule, but DNA *in vivo* often has a closed structure without any free ends. In bacteria and viruses, DNA is often circular. In eukaryotes also, large loops of DNA are found in such a way that each loop represents part of a circle. This organization puts an additional constraint on

A Comparison Between Right Handed (R) and RL Models

Properties	*Watson and Crick's 'R' model*	*Sasisekharan's 'RL' model*
1. Helix sense	Right handed throughout	Alternating right and left handed segments
2. Rotation/10 bp	360°	0° to 36° (ideally 0°)
3. Conformational flexibility	No	Yes

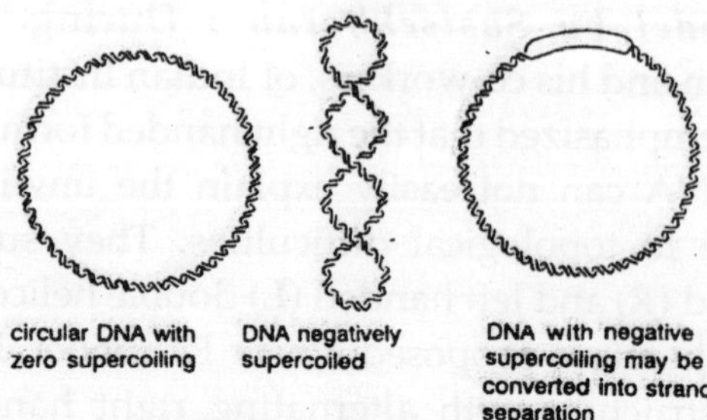

Supercoiling of DNA leading to a twisted duplex, which may be undone by strand separation in a region.

double helical structure, and the DNA becomes supercoiled (Fig.). One supercoil is introduced every time the duplex thread is twisted about its axis. This supercoiling places a DNA molecule under torsion, which can be released by a break in one of the two strands. A DNA molecule without supercoiling is said to be relaxed.

The supercoils may be negative (as found *in vivo),* when they are in a direction opposite to the clockwise turns of right handed DNA. This will actually relieve the torsion and will become underwound or even single stranded in a region. This helps in unwinding of DNA for replication, etc. The positive supercoils lead to overwinding (superhelical), which can be created *in vitro,* but does not occur in nature. Supercoiling is often controlled by some enzymes and is described by parameters like linking number, twisting number and writhing number.

Ribonucleic acid (RNA). Although double stranded RNA is not uncommon, mainly RNA is found as single stranded molecules. Chemical analysis shows the following two main differences in the polynucleotide chains of DNA and RNA: (i) While deoxyribose sugar is found in DNA, in RNA ribose sugar is found. This difference between DNA and RNA is the most important and would distinguish the two under all conditions, (ii) While in DNA, adenine, guanine, thymine and cytosine are the four common bases, in RNA the four bases are adenine, guanine, uracil and cytosine. Since there is a large number of other unusual bases found in RNA as well as in DNA and also since thymine may rarely be found in RNA also, this distinction is not as important as that of sugar molecule. There is yet a third distinction described, i.e., double stranded nature of DNA and single stranded nature of RNA. This distinction is no longer tenable in so far as single stranded DNA (e.g. phage $\phi \times 174$, parvovirus) and double stranded RNA (see Table) are now known in a number of cases.

While DNA is perhaps always genetic in nature, RNA is only rarely genetic in nature. This would mean that although the primary function of storing and carrying the genetic information is rarely also associated with RNA in some organisms, but mainly RNA, found in all organisms, would perform different important functions during protein synthesis. Therefore, two types of RNA are really known, namely genetic RNA and non-genetic RNA. *Genetic RNA.* In most plant viruses, genetic material consists of RNA, although now some plant viruses are known which have DNA. In many bacteriophages also, genetic material is RNA rather than DNA. Viroids and virusoids are other two classes of small RNAs (~350 bases) found in plants. Viroids are subviral pathogens and function independently without encapsidation by a protein coat (e.g. potato spindle tuber viroid or PSTV). Virusoids (also called satellite RNAs) are encapsidated by plant viruses packaged together with a viral genome. Virusoids can not replicate independently, and do so with the help of an associated virus. Such genetic RNA contains information which is normally found in DNA in higher organisms. In other words, RNA has replaced DNA in such cases. The RNA found in these viruses could be single stranded or double stranded (Table).

When RNA is double stranded, it generally follows the same rules of base pairing as in case of DNA. Different viruses having single stranded or double stranded RNA are listed in Table.

Non-genetic RNA : In organisms, where genetic information is contained in, and transmitted through DNA, RNA though present in good quantity, but does not serve as a genetic material. This nor.-genetic RNA is synthesized on DNA template and is of the following three different types.

Different RNA Viruses and the Nature of RNA Associated with them.

Virus	*Type of RNA*
Plant Viruses	
MV	Single stranded
Wound	Double stranded
Animal Viruses	
Influenza virus	Single stranded
Rous sarcoma	Single stranded
Poliomyelitis	Single stranded
Reovirus	Double stranded
Bacteriophages	
MS2, F2, r17	Single stranded

(i) *Messenger RNA (mRNA)* carries genetic information contained in DNA and makes a small fraction (5%-10%) of total RNA present in the cell. This species of RNA is rather short-lived and therefore has a rather rapid turnover. The molecular weight of this RNA varies, usually from 500,000 to 2,000,000.

(ii) *Transfer RNA (tRNA),* also known as soluble RNA (sRNA), makes another small fraction (10%–15%) of RNA. These are the smallest molecules of RNA and

work as adapter molecules for carrying amino acid molecules to the site of protein synthesis. Being smallest in size, these molecules have been most thoroughly investigated.

(iii) *Ribosomal RNA (rRNA)* is the most stable kind of RNA and is associated with the ribosomes, of which this RNA species makes 40%-60% by weight. This species of RNA also makes about 80% of the total RNA in the cell.

Besides the above three major classes of RNA, there is a variety of other non-genetic RNA species found in cells for specific functions. These include the following : (i) Anti-sense RNA also called mic RNA (messenger RNA inhibiting complementary RNA) is synthesized sometimes on the strand complementary to the one used for mRNA synthesis. This is used for regulation of DNA synthesis and gene expression both *in vivo* and *in vitro*. (ii) HnRNA is synthesized from split genes in eukaryotes, and particular class (group 1) of intron

A Comparison of the Primary Structure of Nucleic Acids and Proteins.

Features	*Nucleic Acid*	*Protein*
1. Polymer	Polynucleotide	Polypeptide
2. Basic units	Four types of nucleotides (a) in DNA-pAdR, pTdR, pCdR, pGR (b) in RNA-pAR, pUR, pCR, pGR	Twenty essential amino acids listed in Table; all these amino acids need not to be present in a particular protein.
3. Polarity	A polynucleotide chain has a polarity (5' and 3' end)	A polypeptide would have a free-COOH group at one end and a free-NH_2 group at the other end (C-terminal end and N-terminal ends)
4. Linkage	The nucleotides are linked with the help of 'phosphodiester' bonds (-O-P-O-)	The amino acids are linked with the help of 'peptide' bonds (-CO-NH)

sequences are self-spliced due to ribozyme activity of RNA. (iii) SnRNA (small nuclear RNA), SnoRNA (small nucleolar RNA) and ScRNA (small cytoplasmic RNA) are found in the nucleus, nucleolus and cytoplasm respectively. SnRNAs (U1, U2, U4, U5, U6) are used for the formation of spliceosome leading to splicing of another category (group 2) of intron sequences. (iv) Guide RNA (gRNA), synthesized from minicircle mitochondrial DNA (mtDNA), is used for RNA editing leading to modification of RNA transcribed from DNA template.

Proteins

Different genes in an organism are generally expressed through proteins which are synthesized through the information stored in genes. It is through these proteins that phenotype is eventually determined. There are, of course, some genes which do not express through synthesis of proteins, but are meant for the synthesis of different components of proteins synthesis apparatus. For instance, there are separate genes for different transfer RNA (tRNA) molecules and also for ribosomal RNA (rRNA). These genes do not contain any information for the sequence of amino acids in a protein. In this chapter, a brief introduction only about the structure of proteins will be presented. The details of the mechanism of different steps of protein synthesis will be described later in this book.

Protein Structure : Structure of proteins can be discussed at four levels of organizations: (i) primary structure, (ii) secondary structure, (iii) tertiary structure, and (iv) quaternary structure.

Primary Structure of Proteins : Proteins are polymers of amino acids (described in the previous chapter) and at primary level these can be considered to be polypeptide chains in the same manner, as the nucleic acids are known to be polynucleotides (Table). The polymer (polypeptide) in a protein consists of amino acids linked with the help of peptide

(–CO-NH) bonds. The peptide bond consists of a covalent bond between carboxyl group of one amino acid and amino group of another, shown in Figure. The protein backbone consists of the repeating series of alpha carbons, carboxyl carbons and amino nitrogen.

Secondary Structure of Proteins : A polypeptide chain always has a spatial (in space) organization. The configuration of backbone of polypeptide chain would mean the secondary structure of protein, of which a particular protein may be a part. These secondary structures fall into three main categories: (i) helices (mainly the alpha helix) which can be left handed or right handed (Fig.), (ii) beta strands (backbone is extended or stretched out) or beta sheets (where two or more beta strands are arranged in rows (Fig.) and (iii) turns connecting the helices

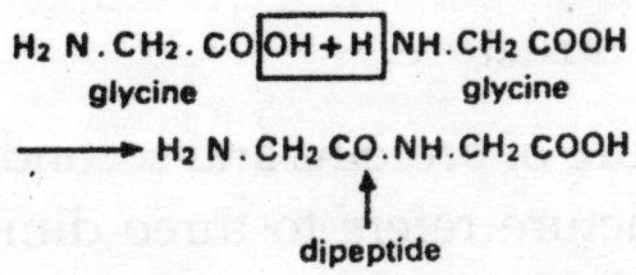

Formation of a peptide bond between two molecules of an amino acid to form a dipeptide.

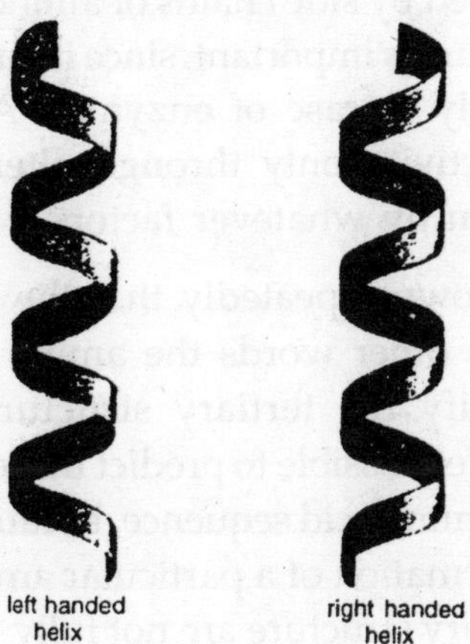

A comparison of a left handed helix and a right handed helix (as a memory device, remember that screws which insert when turned clockwise are right handed and the other is left handed).

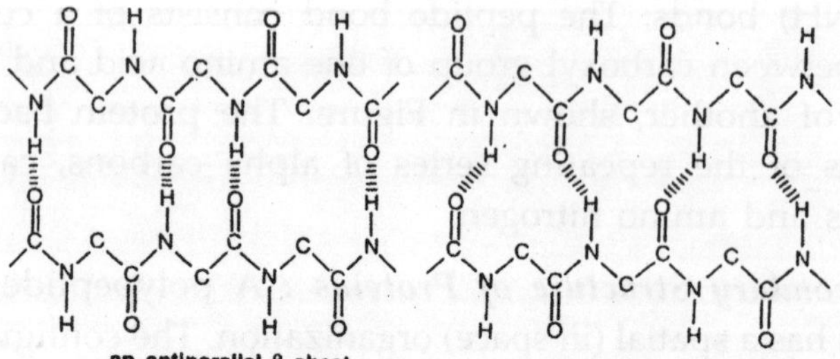

an antiparallel β sheet a parallel β sheet

Two β sheets (an antiparallel and a parallel β sheet)

and strands (beta turns and omega turns). These three secondary elements are shown in Figure. The secondary structures are often stabilized by intramolecular hydrogen bonding. These secondary elements can also combine with one another to form motifs, or super-secondary structures, which assemble in the form of a tertiary structure.

Tertiary structure of proteins and second half of the genetic code. Tertiary structure refers to three dimensional structure of a protein. While the secondary structure is obtained by folding and intramolecular bonding in polypeptide, tertiary structure is stabilized by side chains of amino acids. This three-dimensional structure is important, since it forms sites of specific activity, particularly in case of enzymes. An encyme would lose its specific activity only through alterations in tertiary structure of protein by whatever factors it is brought about.

It has been shown repeatedly that the primary structure of a protein (or in other words the amino acid sequence) is sufficient to specify the tertiary structure of a molecule. However, it is not yet possible to predict the tertiary structure of a protein from its amino acid sequence, because the rules which govern the transformation of a particular amino acid sequence into a specific tertiary structure are not fully understood so far. Hundreds of laboratories around the world are busy working on this problem, which is often described as protein folding problem or also as second half of the genetic code.

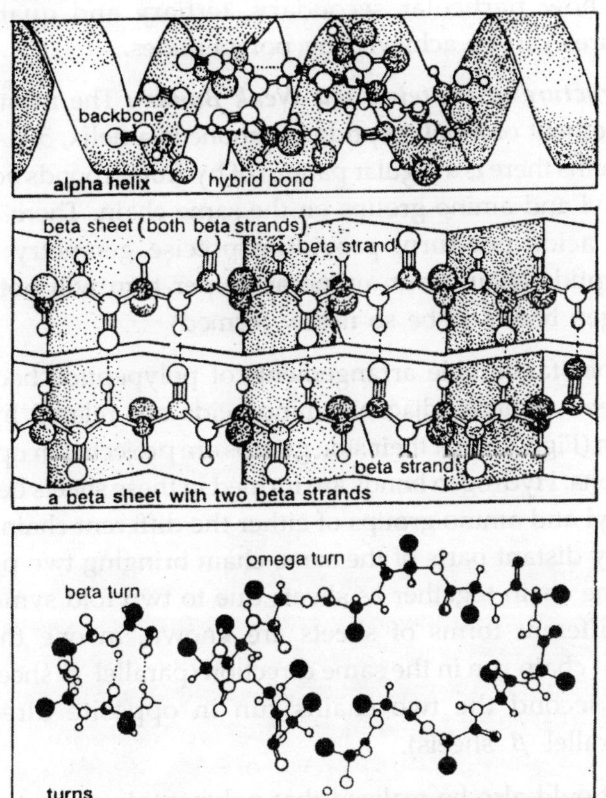

Common secondary structures of proteins (helix, beta sheet and turns).

Quaternary structure of proteins. Primary, secondary and tertiary structures have been discussed with respect to single polypeptide units. However, in a particular protein, more than one polypeptides may be associated to give rise to a complex macromolecule, which is biologically a functional unit. This structure obtained through combination of more than one polypeptides would be quaternary structure. Polypeptide units of this quaternary structure would have their own primary, secondary and tertiary structures. It is important to know that, while the mechanism of synthesis of polypeptides (primary structure) is now known in considerable detail, little is known about how particular secondary, tertiary and quaternary structures can be achieved by polypeptides.

Structure of Proteins and Weak Bonds : The most stable arrangement of a polypeptide backbone is a helix, 5.4 Å long. In proteins there is a regular pattern of hydrogen bonds between carbonyl and amino groups on the same chain. There are 3.6 amino acids per turn, providing precise geometry of the polypeptide chain (four amino acids per turn will not allow hydrogen bonds to be so neatly formed).

Other favourable arrangements of polypeptide backbone are sheets, where adjacent amino acids are related by 180° rotation (Fig.), so that their side groups are projected in opposite directions. Hydrogen bonds are formed in these sheets between carbonyl and amino groups of either the different chains or of the very distant parts of the same chain bringing two parts of the same chain together as sheets due to two fold symmetry. Two different forms of sheets are known, in one the two adjacent chain run in the same direction (parallel β sheet) and in the second the two chains run in opposite directions (antiparallel β sheets).

It should also be realized that polypeptide chains are not mere polymers of identical monomers forming helices or sheets, but invariably have irregular side groups attached to the backbone. This necessitates a compromise between regular helix or sheet backbone and the tendency of the side groups to twist the backbone into much less regular configurations. This maximizes the strength of secondary bonds formed by side groups. In solutions, proteins containing large number of nonpolar side groups tend to be more stable than those containing polar groups. This is because disruption, of molecules with nonpolar group causes much greater loss in free energy than the disruption of molecules with polar groups (disrupted polar groups can make hydrogen bonds with water).

Almost 50% of amino acid residues in proteins do have nonpolar side groups that interact with each other rather than with polar side groups or with water molecules. Within a cell, the proteins assume configurations which allow isolation of nonpolar side group to areas where water can not penetrate (hydrophobic areas).

Ramachandran Plots and Protein Folding : Although, one can not yet predict the three dimensional structure of a protein from its amino acid sequence, protein structure can be reduced to a few elements or motifs, which are arranged in different ways in different proteins. Further, each of the motif can be further understood in terms of the conformational possibilities of the amino acids. Several chemical characteristics determine the motifs. For instance, studies have shown that the six atoms of the peptide group should all be in the same plane, a feature described as planarity of the peptide bond (Fig.). However, some degree of rotation is allowed about the bonds, which join the peptide groups to the α-carbon atom. These rotations are designated as torsion angles ϕ and ψ. The values of ϕ and ψ are influenced to different degrees due to steric hindrance from the side chains. The permissible range of ϕ and ψ can be

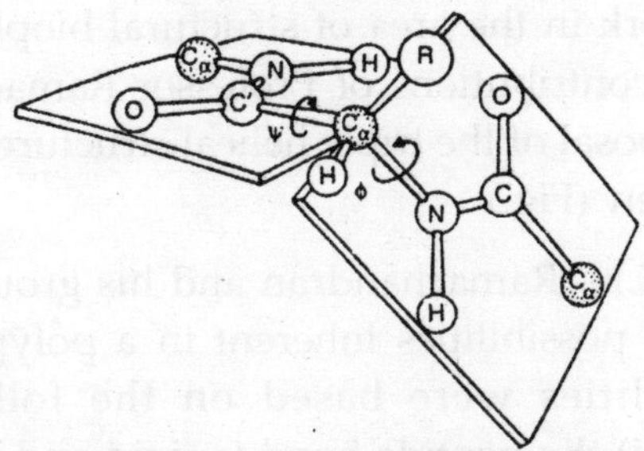

Chemical structure of the peptide bond; partial double bond nature of the peptide bond gives a planar peptide unit, so that the conformation of the protein backbone can be specified in terms of torsion (rotation) angles ϕ and ψ (redrawn from Balasubramanian et al., Concepts in Biotechnology', 1996)

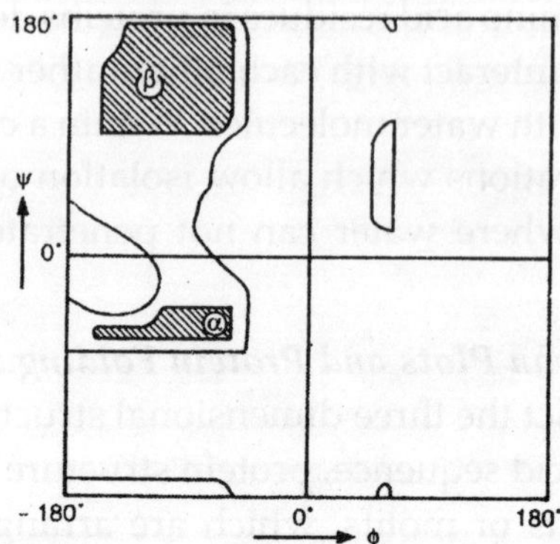

The Ramachandran plot (redrawn from Balasubramanian et al., Concepts in Biotechnology, 1996; see text for details)

plotted against each other to give what are popularly described as Ramachandran plots (Fig.). Some of the values of ϕ and ψ can give regular patterns of atoms, if they are repeated over a number of residues in the protein chain. These patterns are actually the secondary structures. The Ramachandran plots are so called after the name of Professor G.N. Ramachandran, who worked at the Department of Crystallography and Biophysics, Univesity of Madras and later at Indian Institute of Science, Bangalore. He became internationally known for his seminal work in the area of structural biophysics. One of the important contributions of Professor Ramachandran was his correct proposal of the triple helical structure of the fibrous protein, collagen (Fig.).

Professor G.N. Ramachandran and his group studied the conformational possibilities inherent in a polypeptide chain. These possibilities were based on the following valid assumptions ; (i) the peptide bond is rigid and will not allow rotation; (ii) the molecule may be rotated about the other single bonds in the backbone, i.e. the $N-C_\alpha$ bond and $C_\alpha-C$, bonds; (iii) all atoms are represented as hard spheres, there being no chance of inter-penetration (even though we know that atoms are not spheres, but for this purpose they were treated as hard spheres) ; (iv) there are no other attractive or repulsive forces.

A triple helix representing collagen (redrawn from Balasubramanian et al. Concepts in Biotechnology, 1996)

Based on the above assumptions, conformational possibilities of a dipeptide can be represented in the form of a 'Ramachandran map' shown in the Figure. The hatched areas represent the combinations of ϕ and ψ torsion angles allowed for any dipeptide (other than glycine, which has a much larger region allowed, since it is the smallest of the amino acids). These regions expanded a bit, when inter-penetration of the atomic spheres were accepted later. These Ramachandran plots did not change very much over the last more than three decades, during which thousands of proteins were found to have 98% of their residues well within the limits suggested in Ramachandran plots. The values of ϕ and ψ which if repeated in consecutive residues (and thus leading to regular structures), are also marked in the map (α helix and β sheets). Some of these secondary structures are described in terms of ϕ and ψ.

α ***-helix :*** When (ϕ, ψ) values (–57°, – 47°) are repeated, the polypeptide backbone chain of the protein is folded in a right-handed manner around an axis giving it the appearance of a helix with side chains extending out. This α helix is stabilized by hydrogen bonds, which are formed between C–O group of each amino acid and –N–H group of amino acid situated four residues ahead in the linear sequence. Each residue is related to the next one by 1.5Å along the axis and a rotation of 100° about it, thus giving 3.6 amino acid residues per turn and a pitch of 5.4 Å (3.6 × 1.5 = 5.4) and a radius of 2.3Å. (example: keratin protein in hairs or wool).

The β strand and β pleated sheet. When the backbone of a polypeptide chain is extended or stretched out, it forms, what is described as β strand. The extended polypeptide chains can also associate by hydrogen bonds to form sheet like structures. These structures are planar and the polypeptide chains involved may be parallel or antiparallel. These are

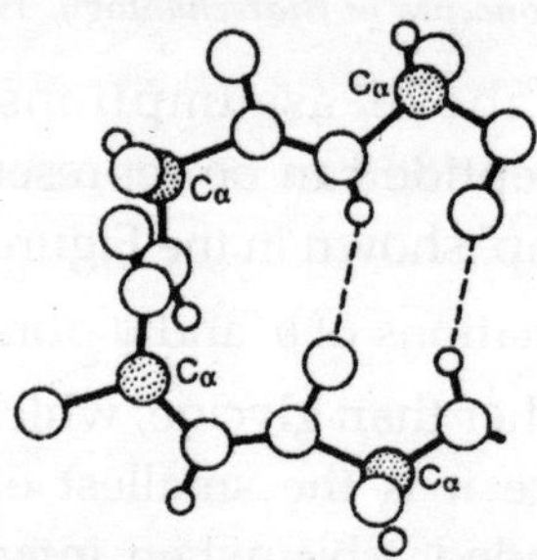

A P-turn conformation (redrawn from Balasubramanian et al., Concepts in Biotechnology, 1996)

called β sheets. The distance between adjacent residues, along the direction of the sheet is 3.5Å. Hydrogen bond between N – H and C – O of different polypeptide chains stabilize the structure. The silk fibre exist in β -shrand or β -sheet, which gives the silk its tensile strength.

β turns (or reverse turns) and ω (omega) turns). The polypeptide chain sometimes reverses its direction, and it does so with the help of a structural motif called β turn or reverse turn or omega turn (Fig.). Such turns are abundantly formed in proteins, particularly in globular proteins. Turns sometimes also connect helices and strands as in the Rossmann fold (Fig.) giving rise to a super-secondary structure.

Protein Conformation Based Inheritance—the Prions : As discussed earlier in this chapter among cellular macromolecules, nucleic acids are known to be the sole hereditary determinants. However, during 1980s and 1990s, cases have been reported, where proteins were shown to determine inheritance through conformational changes. This is popularly described now as "protein conformation based inheritance," although some workers believe that the evidence in favour of this type of inheritance is not unequivocal. This protein conformation based inheritance is exhibited by infectious particles called "prions". The term 'prion' was introduced in 1982 by Stanley B Prusiner of the University of California (San Francisco), who has been awarded the 1997 Nobel Prize in Medicine. The term was used

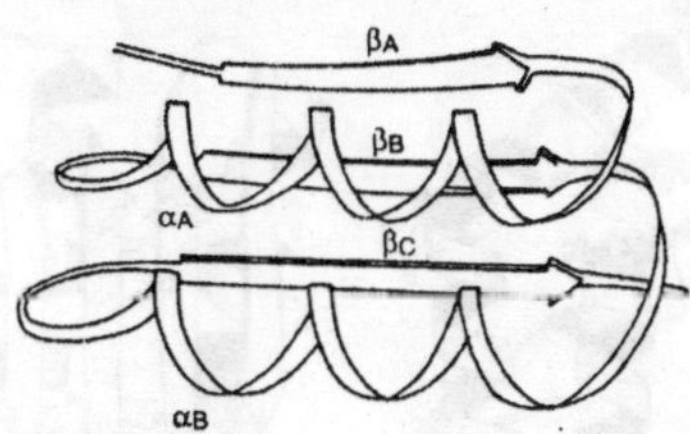

The Rossmann fold, a common .super-secondary structural unit. It alternates between α β strand and an a helix along the chain; the β strands are arranged in a sheet and α helices are held together (redrawn from Balasaburamnian et al., Concepts in Biotechnology, 1996)

to characterize the proteinaceous infectious particles causing a variety of mammalian neurodegenerative diseases which are generally fatal and are referred to as transmissible spongiform encephalopathies (TSEs). These include diseases like Scrapie, CJD (Creutzfeldt-Jakob Disease), GSS (Gertsmann-Straussler-Scheinker Disease) and Kuru. Prions also cause the 'mad cow disease' or 'bovine spongiform encephalopathy' (BSE), which created a panic in Britain in 1996, since the disease was suspected to be caused due to eating BSE-tainted beef. Despite being infectious, prions differ from viruses, viroids, retroviruses and virusoids or satellite viruses in having no nucleic acid components whatsoever (viruses have their own DNA/RNA and proteins; viroids are RNA molecules not encapsidated; virusoids are nucleic acid molecules encapsulated by protein coat derived from other viruses, and the retroviruses carry cellular oncogenes). The absence of nucleic acids and presence of only proteins in these pathogens was initially proved, when it was observed that UV and ionizing radiations (which damage nucleic acids) did not cause any loss of infectivity, while treatments involving modification or hydrolysis of proteins caused loss of infectivity.

The prion protein is designated as PrP. Since it occurs both in the normal and the infected hamsters, apparently it can not

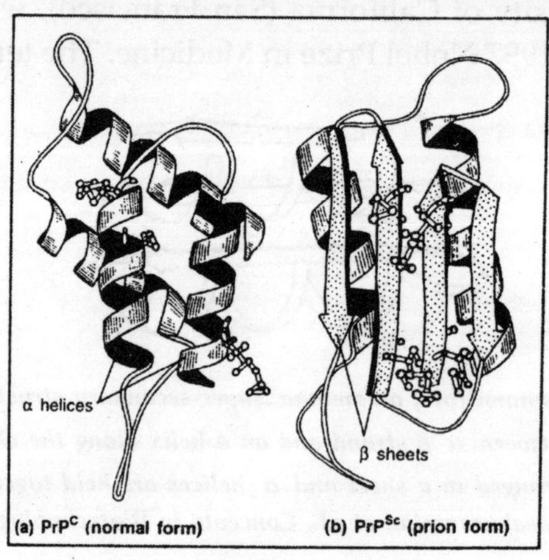

Two conformations of prion protein. PrP^C (rich in α helices) and PrP^{Sc}, (rich in β sheets)

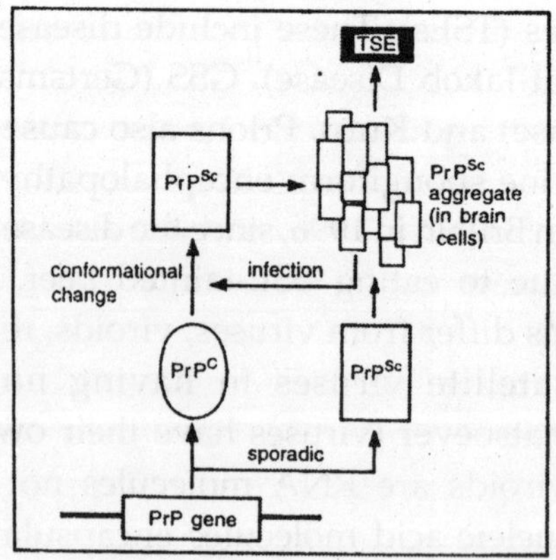

A hypothesis suggesting conformational change in prion protein PrP^C giving rise to PrP^{Sc}, which forms aggregates and fibrils in brain cells causing 'transmissible spongiform encephalopathies' (TSE).

cause the disease. Since the protease resistant core of PrP has a molecular weight of 27-30kDa, it is also known as PrP 27-30. This PrP is only a part of a larger protein PrP^{Sc} with a molecular weight of 33-35 kDa (superscript Sc stands for protein causing scrapie), which causes infection. A subset of PrP is also designated as PrP^{l}, which is protease sensitive and works as a substrate for PrP^{Sc}, the essential component of the infectious prion particle. Physically, PrP^{Sc} differs, from PrP^{C} by its insolubility in detergents, its propensity to aggregate and its relative resistance to proteolysis. While PrP^{c} is rich in α helices, PrP^{Sc} is rich in β-sheets (Fig.). Further, unlike PrP^{C} PrP^{Sc} tends to form insoluble aggregates and fibrils in brain cells (Fig.).

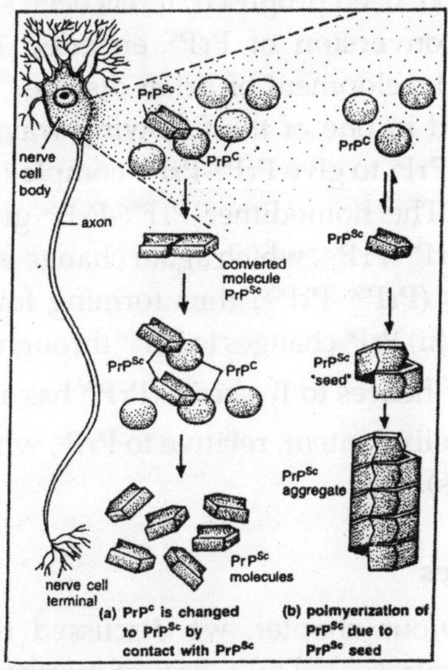

Two models for post-translaticnal conversion of PrP^{c} into PrP^{Sc}.

The prion diseases have also been shown to be hereditary. The PrP gene in Syrian Hamster has two exons, while that of mouse has three exons. However, the open reading frame (ORF) of all the known PrP genes is present in a single exon in each case, so that PrP^{Sc} can not result due to alternative RNA splicing, but should only result due to post translational modification. Several prion diseases are actually caused due to mutations in the PrP genes *(Prn-P)*.

Genetic nature of prion diseases was also demonstrated by using transgenic mice carrying *Prn-P* from Syrian Hamster (SHaPrP) and also by ablation of the *Prn-P* in mice *(Prn-$P^{0/0}$)*. In both these cases it was shown that the de novo synthesis of prion is species specific or transgene specific.

Since there is no evidence available for the presence of scrapie-specific nucleic acid, and since proteins are incapable of undergoing self replication, alternative mechanisms for prion propagation have been proposed. It has been shown that post-translational conversion of PrP^C encoded in PrP gene is obligatory for development of prion disease. This conversion can be achieved in one of the following ways (Fig.) (i) PrP^{Sc} combines with PrP^c to give PrP^C-PrP^{Sc} complex, which changes to PrP^{Sc}–PrP^{Sc}. The homodimer PrP^{Sc}-PrP^{Sc} gives rise to two heterodimers PrP^c–PrP^{Sc}, which again change to two molecules of homodimers (PrP^{Se}–PrP^{Sc}) thus forming four molecules of PrP^{Sc} and so on; (ii) PrP^c changes to PrP^{Sc} through conformational change from α -helices to β -sheets (PrP^{Sc} has a higher β -sheet and lower α -helix content, relative to PrP^c, which has four α -helical domains).

Polysaccharides

In the previous chapter, we discussed the structure of simple sugars represented by a class of micromolecules. These sugars included monosaccharides and oligosaccharides. Polysaccharides represent a class of macromolecules that contain, each at least 10 or more monosaccharide units. They are given names by replacing the ending -ose in the name of their monomers, by -an. For instance, polysaccharides derived from glucose (e.g. cellulose) are glucans. Similarly, polysaccharides derived from fructose are called fructans. A general name for polysaccharides is glycan, because monosaccharides are given the general name glycose. The popular names starch, cellulose, pectin, amylopectin, inulin, chitin and heparin are still used, although they do not conform to the above nomenclature, where all poly saccharides should have names ending with '-an'.

Polysaccharides have been classified into two groups : (i) homopolysaccharides (also called homoglycans) containing only one type of monosaccharide or glycose units and (ii) heteropolysaccharides (also called heteroglycans) containing atleast two different monosaccharides (or glycose) units (Table). Most heteropoly saccharides do not contain more than two types of glycose units. These polysaccharides are formed due to condensation of many molecules of monosaccharides with simultaneous release of water molecules. Because each monosaccharide has several free hydroxyl groups that can be used for condensation with another monosaccharide, the number of possible structures of a polysaccharide involving the same monosaccharide units, can be large. This was illustrated in the previous chapter, where a disaccharide consisting of two glucose molecules were shown to have eleven possible structures. Three hexoses ($C_6H_{12}O_6$) can join in a variety of ways to give thousands of different trisaccharides and so on. In view of this, it is extremely difficult to determine the structure of a polysaccharide, because this will involve the determination of the sites of linkage between every two adjacent monosaccharide units. Consequently, it has been shown to be much more difficult to determine the structure of a polysaccharide than to determine the nucleotide sequence in a nucleic acid molecule or to determine the amino acid sequence in a protein.

Homopolysaccharides (Homoglycans) : As mentioned above, homoglycans will consist of same monosaccharide units (e.g. glucose, fructose, galactose, mannose, xylose, etc.). The best examples of homopolysaccharides are glycogen in animals and starch and cellulose in plants. All of them are composed of only glucose residues. While glycogen and starch are used for storing energy, cellulose makes the extracellular matrix (ECM) or cell wall in plants. Different homopolysaccharides, their repeat units and the linkage type in each case are summarised in Table.

Starch and glycogen. Starch is the reserved food in plant cells and is a polymer of D-glucopyranose units linked by α-1, 4 linkages (occasionally α-1, 6 linkages are found in amylopectin giving it a branched structure). It consists of amylose and amylopectin in the ratio of 1 : 4. While amylose has a linear structure (200-500 glucose units; mol. wt. ~ 1500,000), amylopectin has a branched structure (1000 glucose residues; mol. wt. ~200,000 to 1000,000 (Figs.). If starch is mixed in hot water, only amylose is dissolved and not the amylopectin. Glycogen is the major reserve carbohydrate in animals and is, therefore, sometimes called animal starch. It is mainly found in liver and muscles. The glycogen in the liver supplies glucose to all tissues through the blood, while that in the muscle is used during muscle contraction. Glycogen consists of only glucose units (5000-15,000 in a molecule) linked by α-1, 4 and α-1, 6 linkages giving it a branched structure with a molecular weight ranging from 300,000 to 100 × 10^6.

Different Classes of Polysaccharides.

	Class of pnlysaccharide	*Linkage*
1.	Homopolysaccharides (homoglycans)	
	(a) Glucans (glucose)	
	Starch	α—1, 4
	Glycogen	α—1,4; α—1,6
	Cellulose	β—1,4
	Chitin	β—1,4

Contd...

Class of pnlysaccharide	*Linkage*
(b) Galactans (galactose)	
Agar	—
Pectin	—
Galactan	—
(c) Mannans (mannan)	α—1,6
(d) Xylans (xylose) (hemicellulose xylan)	β—1,4
(e) Fructans e.g., inulin (fructose)	β— 1,2
2. Heterupolysaccharides (heteroglycans)	
(a) Neutral sugars (several monomers)	
(b) Mucopolysaccharides	
Hyaluronic acid (disaccharide)	β—1,4
Chondroitin sulphates (disaccharide)	β—1,4
Heparin (disaccharide)	—
3. Other heteropolysaccharides (sugars linked to other molecules)	
(a) Glycoproteins (sugars linked to peptides)	—
(b) Lipoproteins (sugars linked to lipids)	—

Cellulose **:** It is the most important constituent of the extracellular matrix (cell wall) in plants. It is insoluble in most solvents and thus differ from starch and glycogen. It is a linear polymer with β-1, 4 glycosidic bonds, where a number of linear chains are held together to form microfibrillae. It can be hydrolysed by cellulase found in some snails, termites, and the bacteria found in the rumen of cattle and other ruminants.

Heteropolysaccharides (Heteroglycans) : Heteropoly saccharides consist of more than one monosaccharides. Sometimes, the repeating unit is a disaccharide. Heteropolysaccharides can be of the following two types.

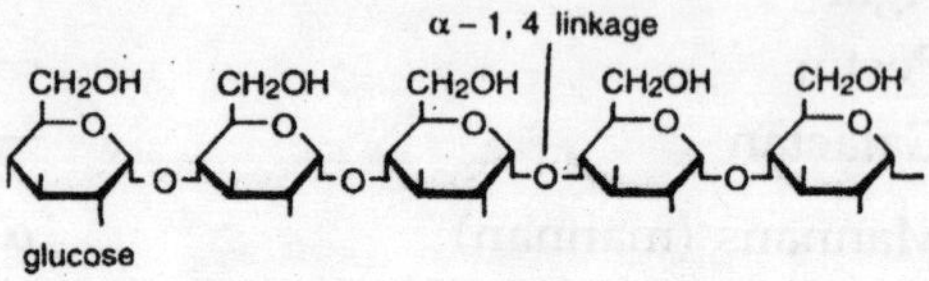

α – 1, 4 linkage
CH2OH CH2OH CH2OH CH2OH CH2OH
glucose

(a) structure of amylose showing α -1,4 linkage

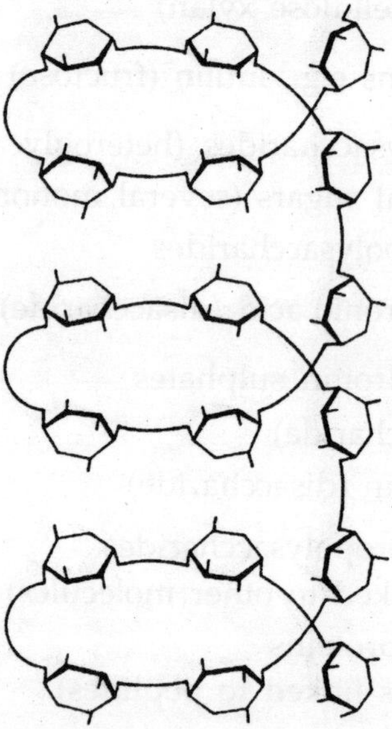

(b) helical coil of amylose on suspension in water

Linear structure of amylose.

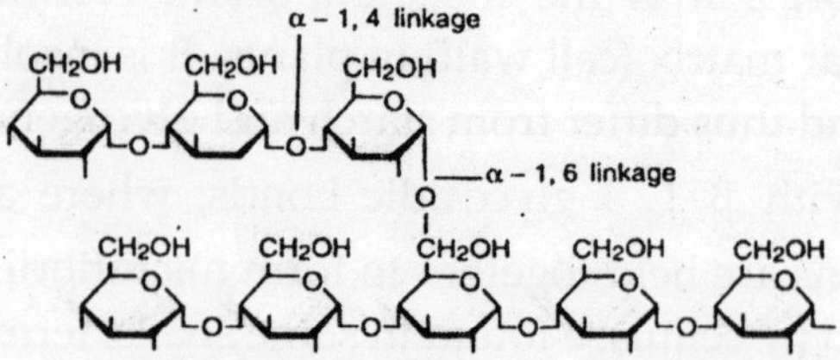

α – 1, 4 linkage
CH2OH CH2OH CH2OH
α – 1, 6 linkage
CH2OH CH2OH CH2OH CH2OH CH2OH

Branched structure of amylopectin.

Neutral Sugars : They include hemicelluloses, gums, mucilages and pectic substances. They contain a number of different sugars (e.g. xylose, arbinose, galactose, rhamnose, mannose, glucuronic acid, etc.) and also some non-sugar compounds.

Mucopolysaccharides **:** They consist of amino sugars, uronic acids and some of their derivatives. Generally they are structural polysaccharides and are found in the connective tissues of animals (eg. blood). Following are the three main examples of mucopolysaccharides : (i) Hyaluronic acid is found in skin, vitreous humour of the eye, the umbilical cord, etc. It is a linear polymer of disaccharide repeat units linked with the help of β-l, 4 glycosidic bonds. In a disaccharide repeat unit there are two monosaccharides-D-glucuronic acid and N-acetyl-D-glucosamine, linked using β-l, 3 bond. (ii) Chondroitin sulphates are found in cornea, cartilage, tendons, skin, heart, saliva, etc. The repeating unit is a disaccharide consisting of glucuronic acid and sulphate ester of N-acetyl galactosamine, linked through a β-l, 3 bond. The disaccharides are linked with each other by β-l, 4 bonds. (iii) Heparin is secreted by most cells in the intestinal mucosa, liver, lung, spleen and kidney. It is an anticoagulant and is used in blood banks for preventing clotting of blood.

Glycoproteins and Glycolipids **:** Polysaccharides are also found associated with proteins and lipids, so that they are then described as glycoproteins and glycolipids. These sugar rich molecules make a sugar rich zone on the surface of plasma membrane, which is described as glycocalyx. A glycoprotein consists of a polysaccharide linked to a polypeptide by a covalent bond. Most of the soluble and membrane bound proteins that are made in the ER are glycoproteins. These are sometimes transported from ER to Golgi apparatus, lysosomes, plasma membrane and extracellular space. The sugar component of these glycoproteins is generally the same in all such glycoproteins. It consists of 14 sugar residues and the association between sugar and protein is facilitated by oligosaccharyl transferase. This polysaccharide consisting of 14 residues is linked to the NH_2 group of an asparagine (N) amino acid in the protein, hence described as N-linked or asparagine-linked (Fig.).

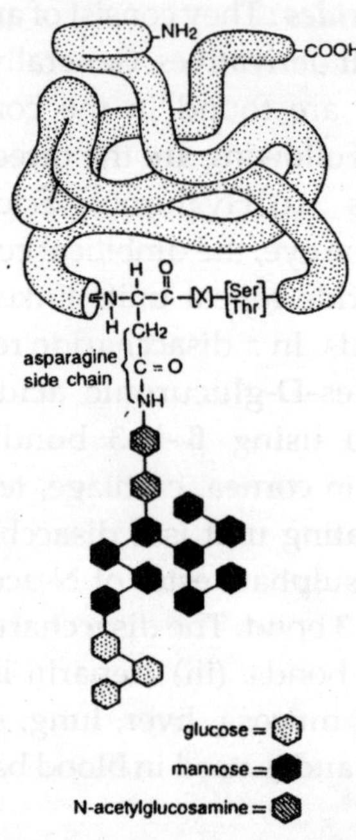

An N-linked or asparagine linked oligosaccharide, (O-linked glycosylation also takes place in some proteins; O stands for serine or threonine), that is added to most proteins in the rough ER membrane (redrawn from Alberts et al. Molecular Biology of the cell, 1994)

Glycolipids are also found in the non-cytoplasmic half of the lipid bilayer in all animal membranes, thus making their distribution asymmetric. These glycolipids work as receptors and also help a cell to bind to its own extracellular matrix (ECM) and to the other cells. The most complex of these glycolipids are gangliosides, which contain oligosaccharides with one or more sialic acid residues. These gangliosides are most abundant in the nerve cells.

2

The Micromolecules

Cells of all living organisms are made up of chemicals, which obey the laws of chemistry and physics. Therefore, in this chapter and the next chapter, we briefly discuss the chemistry of the cell. In this chapter we describe micromolecules or small molecules, which are the basic ingredients of the cell. In the next chapter we will describe macromolecules or giant molecules, which are the polymers of a subset of the small molecules described in this chapter.

The most abundant component of a living cell is water, making about 70% of cell's mass. However, if we disregard water, most micromolecules in a cell are carbon compounds involving other elements like hydrogen, nitrogen and oxygen. The simple rules of covalent bonding between carbon and other elements permit an infinitely large number of compounds, in which groups like methyl ($-CH_3$), hydroxyl ($-OH$), carboxyl ($-COOH$) and amino ($-NH_2$) occur repeatedly in living systems. Thousands of different micromolecules present in the cell make only a small fraction of the total organic matter in a cell (Table). They can be broadly classified into (i) simple sugars, (ii) fatty acids, (iii) amino acids, (iv) nucleotides and (v) lipids. These micro-molecules will be described in this chapter.

The Approximate Chemical Composition of a Bacterial cell (Modified from Alberts* et al., *1994)

Chemical constituent	*Proportion (% total weight)*	*Number of types of each molecule*
1. Water	70.0	1
2. Inorganic ions	1.0	20
3. Sugars and precursors	1.0	250
4. Amino acids and precursors	0.4	100
5. Nucleotides and precursors	0.4	100
6. Fatty acids and precursors	1.0	50
7. Other small molecules	0.2	300
8. Macromolecules	26.0	3000

Sugars (Carbohydrates)

Sugars or carbohydrates are usually classified into three groups : (i) monosaccharides, (ii) oligosaccharides and (iii) polysaccharides. Simplest sugars are monosaccharides with

Common Monosaccharides.

Monosaccharides with general formula	$(CH_2O)_n$	*Examples*	
		Aldehydes	*Ketones*
1 Trioses,	$C_3H_6O_3$	Glyceraldehyde	Dihydroxyacetone
2 Tetroses,	$C_4H_8O_4$	Erythrose	Erythrulose
3 Pentoses,	C_5H10O_5	Deoxyribose	Ribose
4 Hexoses,	$C_6H_{12}O_6$	Glucose	Fructose

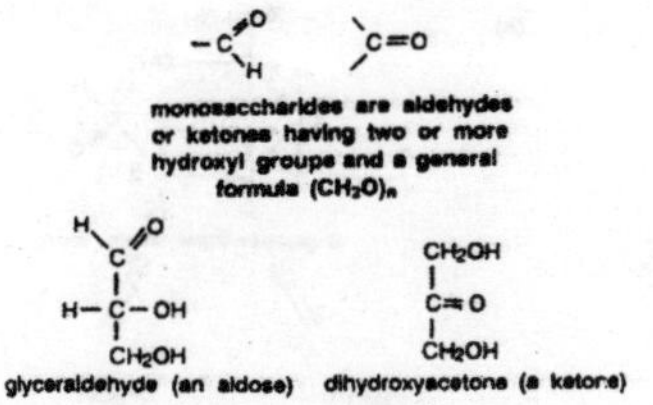

monosaccharides are aldehydes or ketones having two or more hydroxyl groups and a general formula $(CH_2O)_n$

glyceraldehyde (an aldose) dihydroxyacetone (a ketone)

Two simplest monosaccharides, glyceraldehyde and dihydroxyacetone.

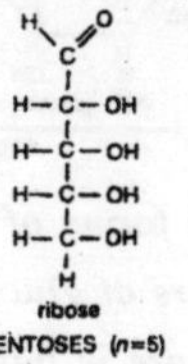

ribose

PENTOSES ($n=5$)

A common pentose-ribose sugar.

the general formula $(CH_2O)_n$, where n is an integer ranging from 3 to 7. The simplest of these monosaccharides are those, with n = 3 giving a molecular formula $C_3H_6O_3$ represented by compounds like glyceraldehyde (an aldehyde) or dihydroxyacetone (a ketone) shown in Figure (Table). Another class of common monosaccharides is represented by pentoses (ribose and deoxyribose), which are used as building blocks for nucleotides and nucleic acids (Fig.). However, the most common monosaccharide found in the cell is glucose with the formula $C_6H_{12}O_6$ and having either an open chain form or a ring form (Fig.). In their open chain form, in addition to a number of

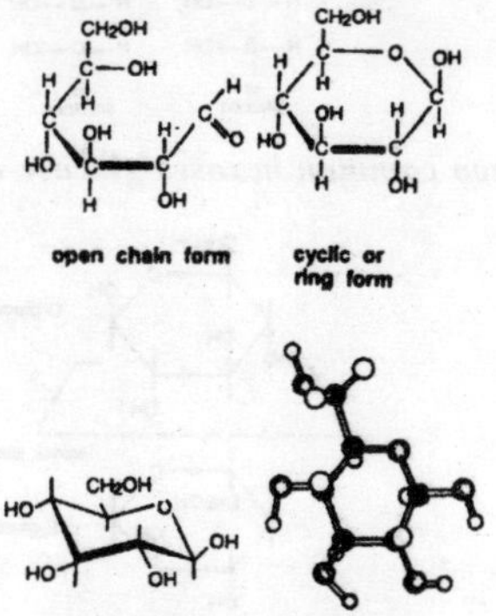

open chain form cyclic or ring form

representation of another cyclic form ball and stick model

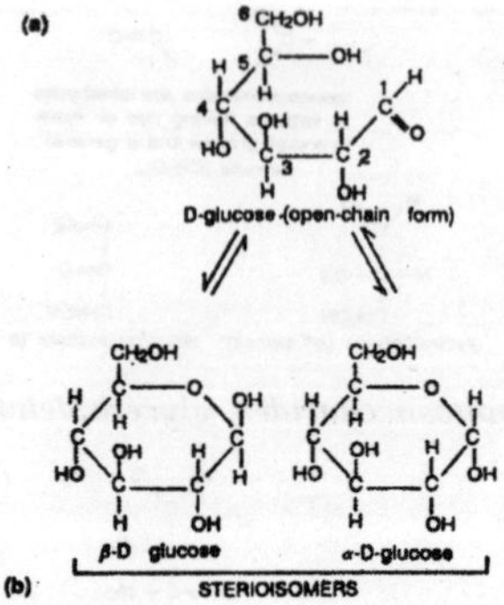

(a) Open and closed-chain forms of glucose (a common hexose sugar); (b) two stereoisomers of glucose (β - D-glucose and α -D-glucose) are in balance with open form.

hydroxyl groups, sugars contain either an aldehyde (H—C = O) group as in glucose or a keto (C=O) group as in fructose (Fig.). The aldehyde or keto group can react with a hydroxyl group in the same molecule to convert open chain into a ring form, which can exist in several isomeric forms (e.g. D-glucose and L-glucose; α -D glucose and β -D glucose; see Fig.). In its ring form, the carbon of the original aldehyde or

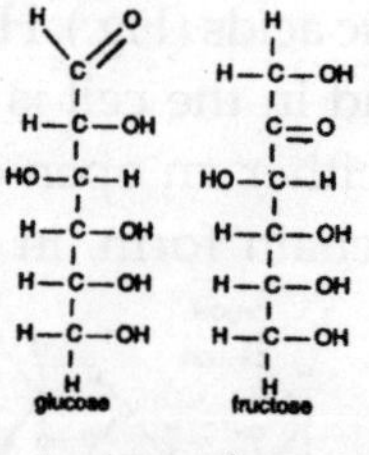

Two common hexoses-glucose and fructose.

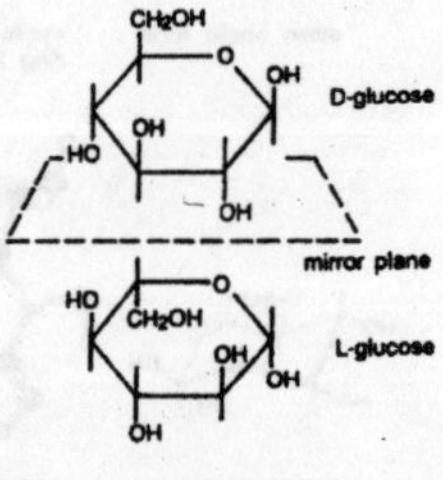

D-glucose and L-glucose are mirror images of each other .

keto group can be recognized as the only carbon atom, that is bonded with two oxygens. The hydroxyl groups of a simple monosaccharide can also be replaced by other groups to give derived monosaccharides (Table). Further, in its ring form, the carbon of the original aldehyde or ketone group in a sugar molecule can become linked to another sugar molecule (through one of its carbon atoms bearing a hydroxyl group) to form a disaccharide. This phenomenon is described as condensation (the reverse process being hydrolysis). The linkage between

Derived Monosaccharides.

Class of derived monosaccharide	*Examples*	
	Monosaccharide	*Corresponding derived monosaccharide*
1. Glycoside	Glucose	Glucoside
	Galactose	Galactosides
2. Sugar phosphates	Glucose	Glucose-1-phosphate
	Glucose	Glucose-6-phoshate
3. Sugar acids	Glucose	Gluconic acid (C-l oxidized to- COOH)
	Glucose	Glucuronic acid (C-6 oxidized to -COOH)
	Vitamin C	Ascorbic acid
4. Amino sugars (hexosamines)	Glucose	Glucosamine, N-acetyl glucosamine
	Galactose	N-acetyl galactosamine
	?	N-acetyl muramic acid,
	?	Neuraminic acid,
	?	Sialic acid

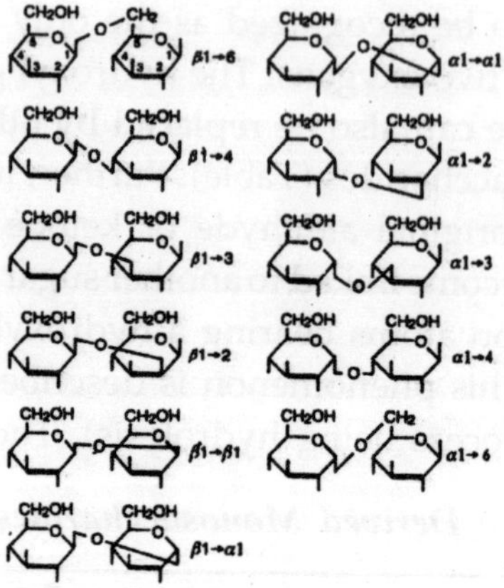

Eleven disaccharides, each consisting of two D-glucose units, and differing only in the type of linkage (shown against each molecule).

Details of Constitution of Some Disaccharides.

Dissachande	*Monosaccharide constituents*	*Glycosidic Linkage*
1. Lactose	D-glucose + D-galactose	β–1,4
2. Maltose	D-glucose + D-glucose	α–1,4
3. Cellobiose	D-glucose + D-glucose	β–1,4
4. Isomaltose	D-glucose + D-glucose	α–1,4
5. Sucrose	D-glucose + D-fructose	α–1, β–2

two monosaccharide sugar molecules is called glycosidic linkage or glycosidic bond, because the hydroxyl group at carbon atom no. 1 (C-1), which is involved in this linkage is described as glycosidic hydroxyl. Depending upon the positions of carbon atoms in two monosaccharide molecules involved in glycosidic linkage, the linkage can be described as 1-4, 1-6 or 1-2. Further, depending upon the α or β forms of each of the two molecules (or α form of one molecule and β form of the other molecule) involved in linkage, the linkages can be described as follows: α-1, 4 linkage; (β-1, 4 linkage; α-1, 6 linkage and α-1, β-2 linkage, etc. Even a simple disaccharide consisting of two glucose residues can exist in eleven different forms (Fig.). Some of the common disaccharides, their monosaccharide constituents and glycosidic linkages are summarized in Table.

Several trisaccharides($C_{18}H_{36}O_{18}$) of biological importance are also known. Raffinose is one such trisaccharide with monosaccharide units, D-glucose, D-fructose and D-galactose. Other trisaccharides are mannotriose, rabinose, rhamminose, gentianose and melezitose. Sugars of higher order are also known. The sugars consisting of fewer sugar residues (generally 2-9) are often collectively described as oligosaccharides.

Sugars having a large number of monosaccharide units (often having atleast 10 units) are known as polysaccharides. These are classified according to their monosaccharide units, e.g. glucans consisting of glucose monomers and galactans consisting of galactose monomers. These polysaccharides represent a class of macromolecules and will be discussed in the next chapter.

Fatty Acids

A fatty acid molecule has two distinct regions: (i) a long hydrocarbon chain, e.g.; CH_3 $(CH_2)_n$ which is hydrophobic (water insoluble) and chemically not very reactive, and (ii) a carboxylic group (–COOH), which ionizes in solution giving COO^-, thus being hydrophilic (water soluble) and chemically reactive. The carboxylic group reacts with a hydroxyl or an amino group of a second molecule to form esters (lipids) or amides respectively. All fatty acid molecules in a cell are usually covalently linked to other molecules by their carboxylic acid group. Hundreds of different kinds of fatty acids exist in the cells and differ firstly in the length of the hydrocarbon chain and secondly in being either saturated having no double bonds (e.g.,palmitic acid), or unsaturated having one double bond (monoenoic acids, e.g., oleic acid), two double bonds (dienoic acids, e.g., linoleic acid), three double bonds (trienoic acids, e.g., linolenic acid) or even many double bonds (polyenoic acids, e.g., arachidonic acid) (Table).

As components of lipids, fatty acids are a valuable source of food, since they can be broken down to produce more than twice as much usable energy as glucose (weight by weight). Fatty acids, in the form of droplets of triglyceride molecules

Some Unsaturated Fatty Acids

Class of unsaturated fatty acid	*Fatty acid*	*Chemical formula showing double bonds*
Monoenoic	Oleic acid	$CH_3(CH_2)_7CH = CH(CH_2)_7COOH$
Dienoic	Linoleic acid	$CH_3(CH_2)_4CH{=}CH{-}CH_2{-}CH = CH(CH_2)_7COOH$
Trienoic	Linolenic acid	$CH_3 - CH_2 - CH = CH - CH_2 - CH =$ $CH - CH_2 - CH = CH(CH_2)_7COOH$
Tetraenoic	Arachidonic acid	$CH_3(CH_2)_4CH = CH.CH_2.CH =$ $CHCH_2CH = CHCH_2CH{=}CH(CH_2)_3COOH$

are stored in the cytoplasm of many cells and represent animal fats. When required, fatty acids are released from these triglycerides and are broken down into two-carbon units to release energy. These two-carbon units, present in the form of acetyl coenzyme A (acetyl CoA), are further degraded in various energy yielding reactions.

Fatty acids are also found as phospholipids and glycolipids forming an important component of cell membranes (see later for structure of phospholipids).

Amino Acids

All amino acids contain a carboxylic group and an amino group, both linked to a single carbon atom called the α-carbon (Fig.). They are the building blocks of proteins, which are a class of macromolecules consisting of amino acids joined by peptide bonds. A peptide bond is formed between the carboxylic group of one amino acid and the amino group of the adjoining amino acid (Fig.).

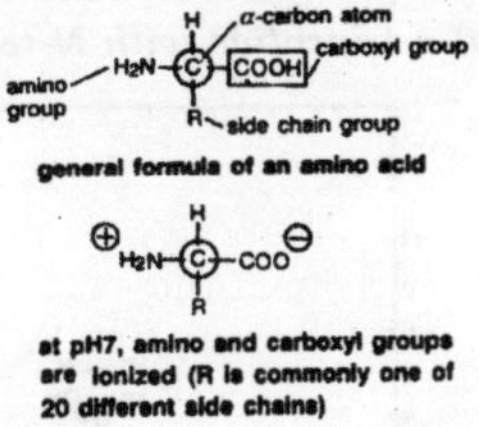

α-carbon atom
carboxyl group
amino group
side chain group
general formula of an amino acid

at pH7, amino and carboxyl groups are ionized (R is commonly one of 20 different side chains)

The general formula of an amino acid showing the positions of amino (-NH2) group, carboxyl group (-COOH), α -carbon atom and a side chain.

Although, there can be many different kinds of amino acids, their number approaching 100, only 20 of them are commonly found in proteins in all classes of organisms ranging from bacteria to higher plants and animals. Different amino acids have an $-NH_2$ group, a-COOH group and an a-carbon in common, but differ with respect to the structure of side chain attached to the α -carbon. These 20 amino acids occurring in proteins are so selected during evolution, that they provide considerable chemical versatility. Five (aspartic acid, glutamic acid, histidine, lysine, arginine) of these 20 amino acids carry each a side chain that can carry a charge, but other 15 are uncharged, though reactive in various specific ways. The charge on the amino acids depends on pH of the solution. In aqueous solution, each of the above five amino acids lose H^+ to form (b) a tripeptide (HIS-CYS-VAL) showing N-terminus (always written to the left and C-terminus always written to the right) a negatively charged ion (e.g. aspartate, glutamate, etc.). In this respect amines are comparable and become positively charged two amino acid molecules join by an amide linkage (—CO—NH—);

the four atoms in the box form a rigid planar unit providing no freedom of rotation about the C—N bond

(a) formation of a dipeptide

amino or N-terminus

carboxyl or C-terminus

these two single bonds on either side of the rigid peptide unit, exhibit a high degree of rotational freedom

(a) Formation of a peptide bond between two amino acids giving a dipeptide and (b) a tripeptide with N-terminus and C-terminus.

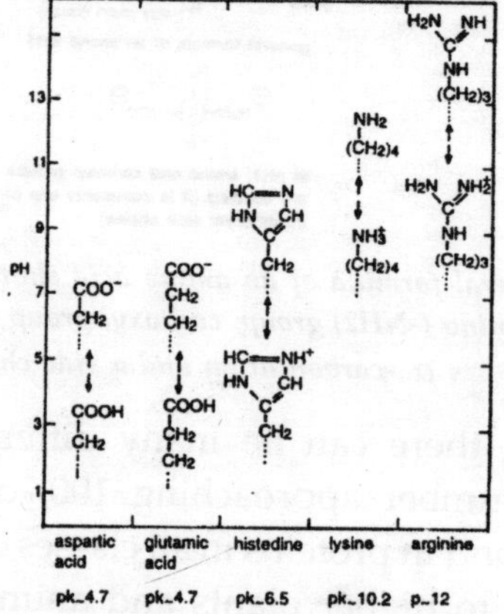

The dependence of the charge on the side chain of an amino acid on the pH; the carboxylic group of the acid loses H⁺ in aqueous solution and degree of this dissociation depends on pH thus controlling the charge; the pH at which exactly half of the carboxylic acid is charged in an amino acid, is known as its pK (redrawn from Alberts et al.: Molecular Biology of the Cell, 1994).

by accepting a H^+. In a solution, both charged and uncharged forms of an amino acid may be found, their relative proportion depending on the pH of the solution. For instance, at high pH, an amino acid is charged and an amine is uncharged and at low pH the situation is reverse. The pH at which half of the carboxylic acid or amine residues are charged is known as the pK of that amino acid side chain (Fig.). In the cell, pH is close to 7 so that all carboxylic acid and amine residues are in their fully charged form.

Abbreviations for 20 Essential Amino Acids

Amino acid	*Abbreviations*	
	Three letter	*One letter*
Alanine	Ala	A
Cysteine	Cys	C
Aspartic acid	Asp	D
Glutamic acid	Glu	E
Phenylalanine	Phe	F
Glycine	Gly	G
Histidine	His	H
Isoleucine	Ile	I
Lysine	Lys	K
Leucine	Leu	L
Methionine	Met	M
Asparagine	Asn	N
Proline	Pro	P
Glutamine	Gln	Q
Arginine	Arg	R
Serine	Ser	S
Threonine	Thr	T
Valine	Val	V
Tryptophan	Trp	W
Tyrosine	Tyr	Y

The amino acids are grouped according to whether their side clains are (i) acidic, (ii) basic, (iii) uncharged polar and (iv) nonpolar (Fig). Amino acids with acidic or basic side chains are very polar and are always found on the outside of protein molecules. The amino acids with uncharged polar side chains are hydrophilic and they are also found on the outside of proteins, but those with non-polar side chains are found inside the protein molecules. Paired cysteines also allow formation of disulphide bonds during formation of proteins. The properties and functions of proteins actually depend on the properties of amino acid side chains found in their constituent amino acids. The 20 amino acids found in proteins have been each given both a three letter and one letter abbreviations. Of these 20 amino acids, nine are essential amino acids (Table).

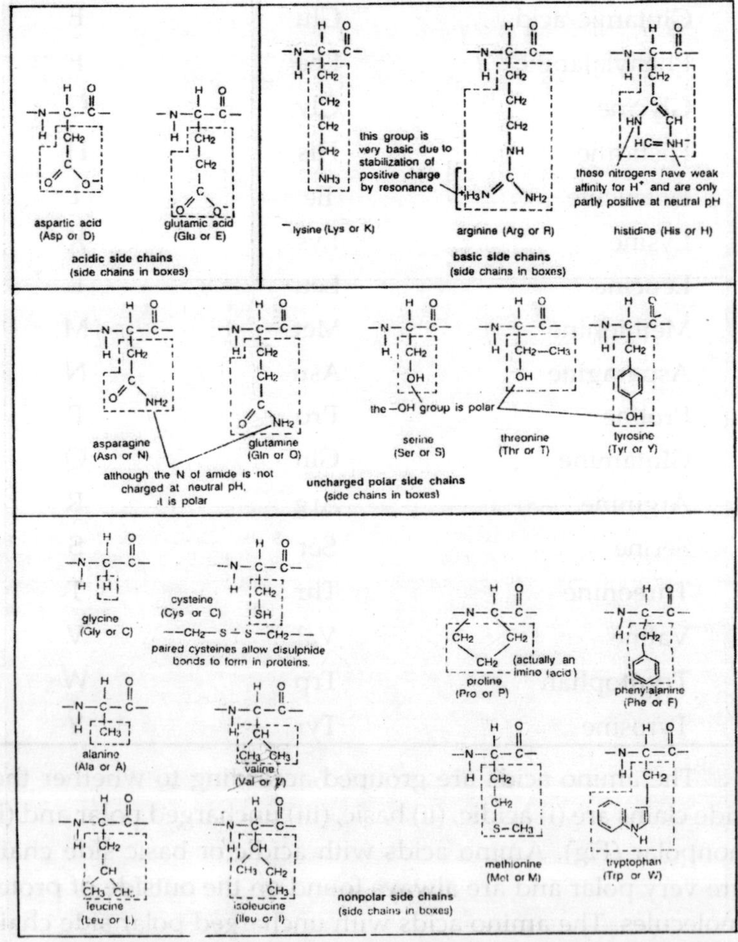

Four families of amino acids, grouped according to acidic and basic nature of the side chain and the charge carried by this side chain.

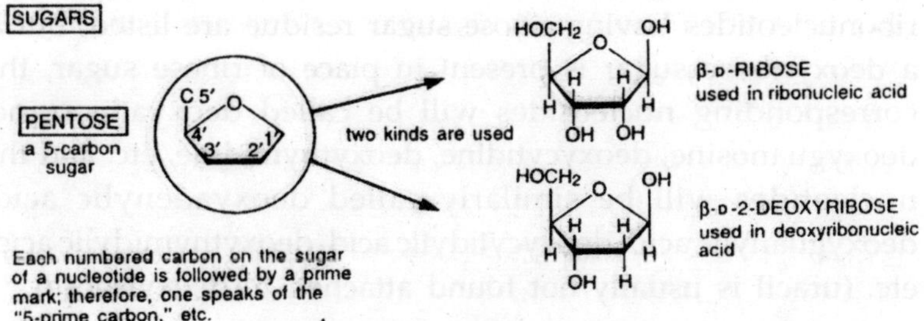

Two pentose sugars, (β -D-ribose found in RNA and β -D-2-deoxyribose found in DNA.

Bases, Nucleosides and Nucleotides.

Class of Bases	*Bases*	*Nucleosides*	*Nucleotides*	*Abbreviations*
I. Purine bases	1. Adenine	Adenosine	Adenylic acid or adenylate	A
	2. Guanine	Guanosine	Guanylic acid or guanylate	G
II. Pyrimidine bases	1. Cytosine	Cytidine	Cytidylic acid or cytidylate	C
	2. Uracil	Uridine	Uridylic acid or uridylate	U
	3. Thymine	Thymidine	Thymidylic acid or thymidylate	T

Nucleotides

A nucleotide consists of a nucleoside, to which a phosphate group is attached, and a nucleoside in its turn consists of a base (so called because it can combine with H^+ in acidic solution) linked to a five-carbon sugar (also called pentose), which may be ribose or deoxyribose (Fig.). The nucleosides or nucleotides can be of five types and their names are derived from the bases they contain (Table).

In Table, the nomenclature of only ribonucleosides and

ribonucleotides having ribose sugar residue are listed, but if a deoxyribose sugar is present in place of ribose sugar, the corresponding nucleosides will be called deoxyadenosine, deoxyguanosine, deoxycytidine, deoxythymidine, etc. and the nucleotides will be similarly called deoxyadenylic acid, deoxyguanylic acid, deoxycytidylic acid, deoxythymidylic acid, etc. (uracil is usually not found attached to deoxyribose).

Nucleotides are primarily used as building blocks of nucleic acids (RNA and DNA), which are polymers of nucleotides, each linked by two ester bonds (phosphodiester linkage) to another nucleotide (Fig.). There are two main types of nucleic acids, which differ in the type of sugar, which forms the backbone : (i) ribose sugar is found in ribonucleic acid or RNA, which contains the bases A,U,G and C and (ii) deoxyribose sugar (where oxygen at position 2 is missing) is found in DNA. The sequence of bases in DNA or RNA represents the genetic information, it contains.

Nucleotides have many functions other than the storage of genetic information. These functions include the following:

(i) They carry chemical energy in their easily hydrolyzed acid-anhydride bonds as in adenesine tri-phosphate or ATP (Fig.). The triphosphate ester of ATP participates in the transfer of energy in hundreds of individual cellular reactions. The terminal phosphate during the production of ATP is added using energy from oxidation of food stuffs and this phosphate can be split off readily by hydrolysis to release energy for other reactions within the cell.

(ii) Nucleotides also combine with other groups to form enzymes as in coenzyme A or CoA (see Fig.).

(iii) Nucleotides like cyclic adenosine monophosphate (cyclic AMP or cAMP) are also used as specific signaling molecules in the cell (Fig.).

Two nucleotides joining by a phosphodiester bond to form a dinucleotide.

Lipids

Lipids are naturally occurring biomolecules comprising a large and diverse group of compounds that are saponifiable esters of long chain (saturated or unsaturated) fatty acids. They are soluble in organic solvents like chloroform, ether and benzene, but are poorly soluble in water. However, sphingomyelins and cerebrosides are insoluble in ether, while phosphatides (glycero-phospholipids) are insoluble in acetone. The lipids have polar (ionic) as well as nonpolar (non-ionic) groups, so that hydrophobic nature of hydrocarbons is associated with hydrophilic nature of other groups like carboxylic group.

All lipids are not esters of fatty acids. For instance, in cerebrosides, the fatty acids's moiety is an amide and in many other lipids there are ether linkages (e.g. ether lipids or plasmogens). Following are the main biological functions of lipids : (i) they are structural components of all living systems; (ii) they are major reserve source of energy, giving 9.7kcal/g, twice that obtained from carbohydrates - this is used for muscular work and maintenance of body temperature; (iii) they are carriers of fat-soluble vitamins A, D, E and K; (iv) some of the hormones are also lipids or their derivatives; (v) they provide essential fatty acids such as linoleic and linolenic acids; some of these fatty acids (e.g. linoleic acid) are precursors for biosynthesis of prostaglandins, which are essential for maintaining several functions of the body.

Lipids can be either simple lipids,which yield monohydric or trihydric alcohol and fatty acid on hydrolysis (due to lipases) or they may be complex lipids, which contain phosphate, various sugars and nitrogenous compounds such as sphingosine, ethanolamine and serine. A summary of the classification of lipids is presented in Figure.

Fats and Waxes

Fats and waxes are esters of fatty acids with alcohols; while fats are esters of fatty acids with glycerol (e.g. triglycerides), waxes are esters of fatty acids with higher chain length aliphatic monohydric alcohols (e.g. lanonin, secreted in the skin of most fur bearing animals; spermaceti, a wax found in the sperm they carry chemical energy in their easily hydrolysed acid- anhydride bonds (e.g. ATP)

they combine with other groups to form enzymes (e.g. coenzyme A or CoA)

they are used as specific signaling molecules in the cell (e.g. cyclic AMP or cAMP)

Three nucleotides, which are not parts of nucleic acids, and have functions other than storage of genetic information.

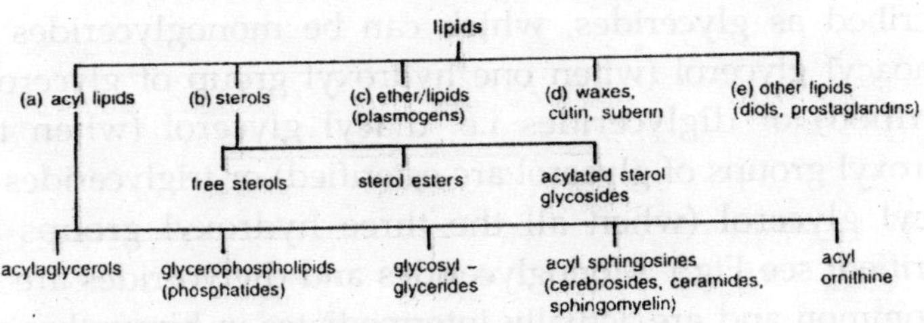

A classification of lipids, showing five main groups and their further division into sub-groups.

whale as a source of stored energy, and myricyl palmitate secreted by honey bee = Bee wax). Waxes also serve as protective coatings on fruit leaves. Esters of fatty acids with glycerol are

α CH_2OH β $CHOH$ CH_2OH α'	CH_2OR $CHOH$ CH_2OH	CH_2OR $CHOH$ CH_2OR	CH_2OR $CHOR$ CH_2OR
glycerol	monoacyl glycerol	diacyl glycerol	triacyl glycerol

R = fatty acid residue = $-OCO(CH_2)_nCH_3$

Derivation of three types of acylglycerols due to esterification of glycerol by one, two or three fatty acid molecules (denoted as R).

$$\begin{array}{c} CH_2OR \\ | \\ CHOR \\ | \\ CH_2OR \end{array} \xrightarrow{\text{lipase}} \begin{array}{c} CH_2OH \\ | \\ CHOH \\ | \\ CH_2OH \end{array} + \begin{array}{c} 3\,RH \\ \text{fatty acid} \end{array}$$

a triglyceride — glycerol

$R = CH_3(CH_2)_nCOO^-$

Hydrolysis of a triglyceride to yield three fatty acid molecules and a molecule of glycerol.

$$\begin{array}{c} CH_2OR \\ | \\ CHOR \\ | \\ CH_2O\,(P) \end{array}$$

3sn-phosphatidic acid

A phosphatidic acid molecule, which is glycerophosphoric acid esterified with two molecules of fatty acid.

described as glycerides, which can be monoglycerides i.e. monoacyl glycerol (when one hydroxyl group of glycerol is esterified) or diglycerides i.e. diacyl glycerol (when two hydroxyl groups of glycerol are esterified) or triglycerides i.e. triacyl glycerol (when all the three hydroxyl groups are esterified; see Fig.). Monoglycerides and diglycerides are not so common and are actually intermediates in biosynthesis of triglycerides, which are the common neutral fats, found as main energy storing compounds in the body. A triglyceride on hydrolysis with the help of lipases yields fatty acids and glycerol (Fig.). When molecules of the same fatty acid are involved at all the three ester bonds, triglyceride is described as simple, but when they differ, the triglyceride is described as mixed.

Glycerophospholipids (Phosphatides) : Glycerophospholipids or phosphatides, as the word indicates, are phospholipids (an alcohol + fatty acid + phosphoric acid), which are derivatives of glycerophosphatidic acid (glycerol having two fatty acid groups and a phosphoric acid group; see Fig.). These are widely distributed in animal and plant tissues. The phosphatides differ in the nature of their fatty acids and in the nature of the alcohol (e.g. aminoalcohols, inositols), which is esterified to phosphoric acid group of phosphatidic acid. Four of these phosphatides are shown in Figure of these phosphatides, lecithins are the most abundant and phosphatidyl inositols are frequently used in signal transduction pathways.

3 sn–phosphatidyl choline (lecithin)	3 sn–phosphatidyl ethanolamine	3 sn–phosphatidyl serine	3 sn–phosphatidyl inositol
CH_2OR	CH_2OR	CH_2OR	CH_2OR
$CHOR$	$CHOR$	$CHOR$	$CHOR$
$CH_2OP\text{–}(CH_2)_2\text{–}N\text{–}(CH_3)_3$	$CH_2O\,(P)\text{–}(CH_2)_2\text{–}NH_3^+$	$CH_2O\,(P)\text{–}CH_2\text{–}CH(NH_3^+)(COO^-)$	$CH_2O\,(P)$-inositol

Four different phosphatides derived due to esterification of the phosphoric acid group of the phosphatidic acid, by different groups (choline, ethanolamine, serine, inositol).

Glycoglycerides : Glycoglycerides are lipids with a carbohydrate moiety attached to sn-3 position of glycerol and are mainly found in photosynthetic membranes of higher plants and cyanobacteria. In small quantities they are also found in fungi and bacteria. Common examples are diacylgalactosyl glycerol and diacyl digalactosyl glycerol (Fig.).

diacylgalactosyl glycerol	diacyldigalactosyl glycerol
CH_2OR	CH_2OR
$CHOR$	$CHOR$
CH_2O-galactose	CH_2O-(galactose)$_2$

Diacylgalactosyl glycerol and diacyldigalactosyl glycerol where two hydroxyl groups of glycerol condense with acyl groups and the third is condensed with one or two galactose molecules.

Acyl Sphingosines, Sphingoglycolipids and Sphingo-Phospholipids : Acyl sphingosines and sphingophospholipids are fatty acid derivatives of sphingosine (Fig.), and are widely distributed in plants and animals respectively. Fatty acid residues are usually saturated and each contains a long carbon chain (e.g. lignoceric acid). They include ceramides (acyl sphingosines) cerebroside (ceramides with carbohydrate group; sphingogly-colipids) and sphingomyelines (sphingophospholipids).

sphingosine	dihydroxy-sphingosine
$CH{=}CH\,(CH_2)_{12}\,CH_3$	$(CH_2)_{14}.CH_3$
$CHOH$	$CHOH$
$CH{-}NH_2$	$CH{-}NH_2$
CH_2OH	CH_2OH

A sphingosine molecule, and a dihydroxy-sphingosine molecule.

Steroids and Sterols : Unlike true lipids and phospholipids, steroids and sterols do not contain fatty acids. In animals, steroids include cholesterol, hormone secreted by testis, ovary and adrenal cortex (except relaxin Table); vitamin D precursors,

Hormones Secreted by Testis, Ovary and Adrenal Cortex, with their Characteristics.

Hormones	*Number of C atoms*	*Structure*
1. Male hormones		No side chain at
(a) testosterone	19	C-17; oxo - or
(b) androsteron		hydroxy groups at C-3 and C-17
2. Female hormones		
(a) oestrogens (oesterone)	18	A benzenoid ring; hydroxyl group at C-3
(b) progesterone	21	A two-carbon side chain at C-17

Contd...

Hormones	*Number of C atoms*	*Structure*
3. Adrenal cortex hormones		
(a) glucocorticoids (e.g. cortisol)	21	A two-carbon side chain at C-17;
(b) oxo-or mineralo-corticoids (e.g. 11 -dexycorlisol)		oxo/hydroxy groups at C—3, C-11, C-17

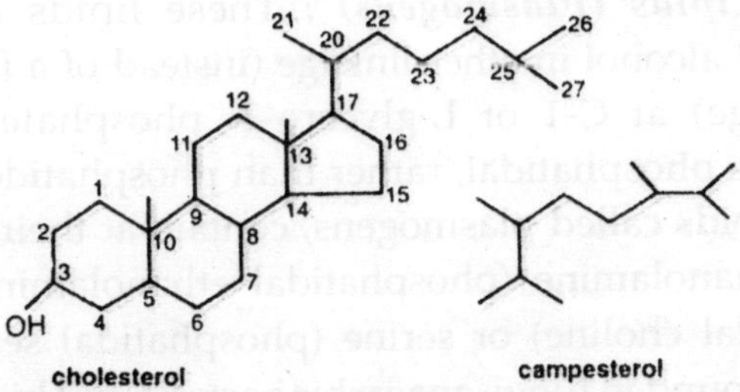

Structure of a cholesterol molecule and a campesterol molecule.

bile acids (catabolic products of cholesterol) and bile salts. In plants, sterols are found either in free form or as sterol esters, acylated sterol glycosides or sterol-glycosides.

Steroids are derived from squalene, a 30-carbon compound, and contain a nucleus (called cyclopentano perhydrophenanthrene nucleus) consisting of three 6-membered carbon rings (A,B and C) and one 5-membered ring (D), with all the rings fully saturated. The carbon atoms of the nucleus are numbered 1 to 17. Sterols are steroids with a hyydroxyl group at carbon 3, and a carbon 17 side chain containing about 10 carbon atoms (e.g. cholesterol; Fig.).

Cholesterol is found in cell membranes (of many organisms), blood and bile. It is also a precursor of several hormones including cortisones testosterone, estrogen, progesterone and other steroid hormones. Ergosterol is precursor of vitamin D.

Free sterols found in plants include sitosterol, stigmasterol and campesterol, which differ in degree of substitution and unsaturation.

Sterol esters are also found in plants and involve esterification of the 3-OH group with a fatty acid (e.g. palmitoyl sterol). Acylated sterol glycosides are storage compounds for sterol and represent its transportable form; they are sterol derivatives, where 3-OH group forms a glycosidic linkage with sugar (usually glucose) and 6th position is esterified with a fatty acid (predominantly palmitic acid/linoleic acid).

Ether Lipids (Plasmogens) : These lipids contain an unsaturated alcohol in ether linkage (instead of a fatty acid in ester linkage) at C-1 of L-glycero-3- phosphate. They are described as phosphatidal, rather than phosphatides. Some of the ether lipids called plasmogens, contain at their phosphate end, an ethanolamine (phosphatidal ethanolamine), choline (phosphatidal choline) or serine (phosphatidal serine). They are mainly found in fungi, anaerobic bacteria and higher plants.

Prostaglandins : They are cydopentoic acids, which are cyclic and hydroxylated derivatives of polyunsaturated fatty acids like arachidonic acid. The different prostaglandins (e.g. PG_E, PG_F and PG_H) differ from each other in the substituted cyclopentane ring and R_1 and R_2 groups (Fig.). Prostaglandins were first discovered in human seminal fluid (hence this name), testis, placenta, uterus, stomach, lung, brain and heart.

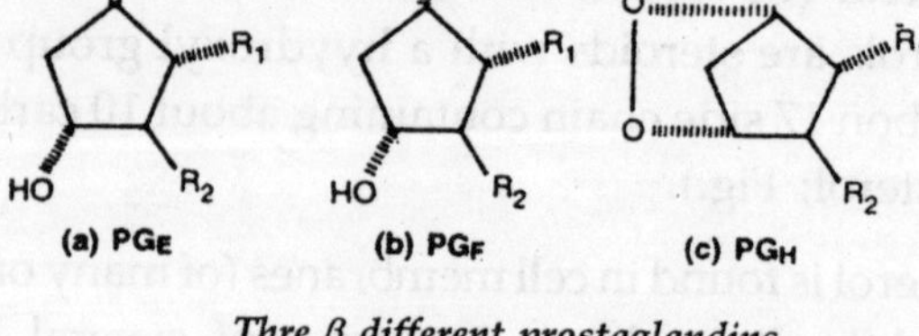

O OH O R₁ R₁ R₁ HO R₂ HO R₂ O R₂

(a) PG_E (b) PG_F (c) PG_H

Thre β different prostaglandins.

3

Animal Genomics

During the last quarter of the 20th century (1975-2000), remarkable progress was witnessed in the field of molecular biology and genetic engineering. Some of the tools used and the progress made in these areas have been briefly discussed in earlier chapters of this book. However, most of this progress was made on the basis of studies involving individual genes or individual DNA fragments/ sequences, rather than the whole genome of an organism. This was a serious limitation and did not allow study of many genes due to non-availability of mutant alleles. During 1980-2000, following two significant developments allowed the study of whole genomes rather than individual genes:

(i) In early 1980s and later, a variety of molecular markers became available, which allowed the construction of whole genome molecular maps;

(ii) Later in mid-1990s, due to the developments of high throughout approaches, it became possible to sequence the whole genomes of individual organisms. This field of research where whole genomes are used for study was described as genomics. Significant progress in the construction of whole genome molecular maps was quickly made after RFLPs became available in 1980. Similarly, significant progress in the field of whole genome sequencing was quickly made after the whole genome sequence (WGS) of the first living organism (other than viruses), *Haemophilus influenzae,* was published in 1995. The methods used for the study of whole genomes and the progress made in the study of whole genomes (construction of molecular maps and whole genome sequencing) in animals will be described in this chapter.

Molecular Markers : Restriction fragment length polymorphisms (RFLPs). What are RFLPs? RFLPs are polymorphisms in specific regions of genomic DNA sequences, which represent recognition sites of specific restriction enzymes. These polymorphisms are due to base alterations including substitutions, insertions/deletions or rearrangements in DNA segments.

The RFLPs are visualized using molecular probes for hybridization, so that sequence polymorphisms are detected in only those genomic regions, which are homologous to the probe used.

Detection of RFLPs: An RFLP can be detected using the following steps :

(i) DNA is extracted and purified from several individuals, which are the subject of study;

(ii) DNA samples are digested, each separately with the same restriction enzyme,

(iii) Each DNA digest is electrophoresed separately, (independent lanes) but on the same gel slab.

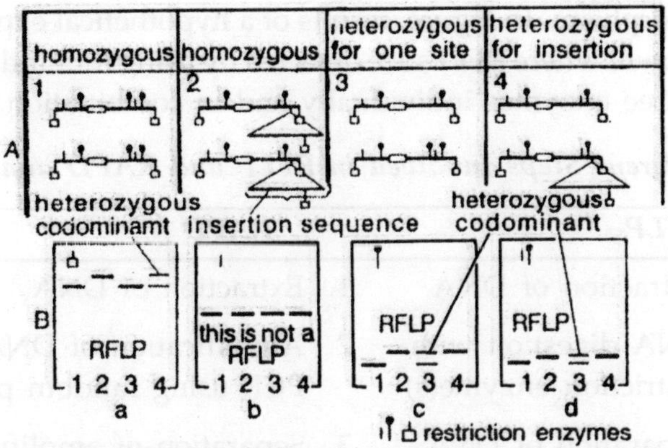

The results of digestion of genomic DNAs of four hypothetical individuals with three enzymes, individually and in combination, followed by hydridization with a specific probe and autoradiography (four individuals are shown as 1-4 in A, 1, 2, being homozygous and 3, being heterozygous for the restriction sites and an insertion; in B, note RFLPs in three of the four cases due to different enzymes.

(iv) The separated DNA fragments are transferred from gel to a nitrocellulose membrane using the technique of Southern blotting.

(v) Southern blots on the membrane are hybridized with a radioactively labelled molecular probe (non-radioactively labelled probes are also often used; see later).

(vi) The hybridized blots are used for exposing an X-ray film, which is developed to obtain autoradiograms. The fragments, which are partially or fully homologous to the probe, can be visualized on the autoradiogram.

RFLP approach is based on the principle of detecting only a few polymorphisms, which are available in sequences related to the probe. The fragments produced by digestion are far too many (with continuous range in size) to be visualized independent of each other, since on electrophoresis they produce a smear on the gel. The use of a probe allows selection of a very small fraction of DNA fragments for the study of polymorphism. In Figure, results of a hypothetical experiment are shown, where RFLPs are detected by using four individuals and three enzymes individually and in combination.

Different Steps Involved in RFLP and RAPD analysis.

RFLP	*RAPD*
1. Extraction of DNA	1. Extraction of DNA
2. DNA digestion with restriction enzyme(s)	2. Amplification of DNA by PCR using random primers
3. Separation of DNA fragments on agarose gel.	3. Separation of amplified DNA on agarose gel.
4. Transfer of DNA fragments-Southern transfers	4. Visualization of markers on the gel and photography of the gel.
5. Hybridization of Southern transfers with probe(s).	
6. Detection of probe using autoradiography or using a non-radioactive system.	

Random Amplified Polymorphic DNAs (RAPDs) : In recent years, polymerase chain reaction (PCR) had an enormous impact on molecular biology research, which is unparalleled. PCR provides selective amplification of a specific DNA segment from an organism, if some information about target DNA is available, so that oligonucleotides may be synthesized to be used as primers at the two ends of the DNA segment to be amplified. If primers with arbitrary sequences (random primers) are used for amplification, DNA segments to be amplified will be selected at random. This will

Properties of Restriction Fragment Length Polymorphisms (RFLPs) and Randomly Amplified Polymorphic DNAs (RAPDs)

	Properties	*RFLPs*	*RAPDs*
1.	Distribution	Ubiquitous	Ubiquitous
2.	Inheritance	Simple Mendelian, stable, co-dominant	Simple Mendelian, stable, dominant/null
3.	Level of polymorphism	High	High
4.	Detect allelic variant?	Yes	No (only null allele detected)
5.	No. of loci detected	1-3	1-10
6.	Part of genome surveyed	Generally low-copy, coding regions	Whole genome
7.	Detectable in all tissues?	Yes	Yes
8.	Technical difficulty	Intermediate	Low
9.	Reliability	High	Intermediate/high (occasionaly spurious) products
10.	Quality of DNA required	Relatively pure	Crude
11.	Quantity of DNA required	2-10 ng	10-100 ng
12.	Use of radioïsotopes?	Yes	No
13.	Type of probes required	Species-specific low copy gDNA or cDNA clones	Random 9- or 10- mer Oligonucleotides as primers
14.	Recurring cost	High	High

provide a truly random sample of DNA markers, which are described as random amplified polymorphic DNAs (RAPDs = pronounced as 'rapids'). RAPDs have several advantages over RFLPs and can be used as their substitute in several studies. The main advantages include the following :

(i) Since same primers with arbitrary sequences can be used for different species, no species-specific probes are needed for different species as required in RFLPs. For instance, primers that provide polymorphic markers in maize, also provided polymorphic markers in soyabean, *Arabidopsis*, humans, *Neurospora*, etc.

(ii) Collection of data using RAPDs proceeds much more quickly than those using RFLPs, since there are fewer steps involved (Table). It has been shown that generation of RAPD data is five times quicker than RFLPs. Important properties of RFLPs and RAPDs are presented in Table for comparison.

RAPD technique may also be combined with RFLPs to increase the efficiency of molecular markers for genetic analysis. A RAPD marker, known to be polymorphic, may be labelled and used as a probe for RFLP analysis, thus eliminating the need for recombinant DNA cloning of probes in bacteria.

VNTRs (Variable Number of Tandem Repeats) : minisatellites, microstallites or SSRs (simple sequence repeats) and ribosomal DNA (rDNA).

It has also been recently emphasized that polyallelic markers will be very useful for mapping both the simple Mendelian traits as well as polygenic traits in segregating populations. The most widespread of these polyallelic markers are the 'variable number of tandem repeat' (VNTR) loci, which are uncovered by locus specific probes or primers and exhibit highly polyallelic fragment length variation. VNTRs include both minisatellites (repeat unit, 11-20 bp long) and microsatellites (repeat unit, 2-6 bp long). Another interesting and common example of VNTRs involves repeat loci associated with rRNA genes in rDNA concentrated at nucleolar organizing regions (NORs) of specific chromosomes of an organism.

Like rRNA genes, most minisatellites are concentrated in proterminal regions of human chromosomes (but not mouse chromosomes) and thus may not be able to provide the desired density of markers. This problem is largely overcome in microsatellite or SSR loci described in the following text.

Microsatellites (SSRs) : In 1989, several laboratories also used PCR to demonstrate a high level of polymorphism or allelic variation in the repeat number (10-50) of SSRs (simple sequence repeats, usually 1-6 bp long), known as 'microsatellites'. Microsatellites occur frequently and randomly everywhere in all eukaryotic DNAs examined, except yeast, and are so named by analogy with the larger minisatellite arrays as above.

The minisatellites, though highly polymorphic, are less common than microsatellites and have larger sequence motifs (upto 1 kbp), making them less amenable to PCR analysis. Microsatellites, on the other hand, are not only amenable to PCR analysis, but are also easy to clone and characterize. They also display considerable stable polymorphism due to variation in the number of repeat units. In view of these attributes, microsatellites are ideal markers for constructing high resolution genetic molecular maps. In 1990, the DNA committee of HGM 10.5 (Human Gene Mapping 10.5) listed 368 microsatellites from human genome (each having motifs of > 4bp and total length > 20bp) available from several databases. They estimated the occurrence of one microsatellite every 6 kbp among 745 kbp of human genomic DNA sequences.

Majority of published microsatellite sequences are actually developed from randomly cloned fragments of total genomic

DNA and characterized. Genomic DNA is digested with a frequent cutting enzyme (Saw3AI, or a combination of *AM, RsaI* and *HaeIII)* and fragments are cloned in sequencing vectors. Genomic libraries generated thus are screened for the presence of microsatellites by hybridization with simple repetitive oligonucleotide probes, e.g. $(AC)_n$, $(AAG)_n$, etc. Positive clones are then sequenced to find out the flanking sequences of SSRs to be used as primers for PCR analysis.

Amplified Fragment Length Polymorphisms (AFLPs) : AFLP is based on PCR amplification of a set of restriction fragments, selected from a pool of fragments that are generated due to digestion of genomic DNA or cDNA with a pair of restriction enzymes. This method generates a large number of fragments ranging in size from a few base pairs to more than 500 base pairs, so that on electrophoresis, as many as 70-100 bands are available from one PCR reaction. It has been estimated that in AFLP, a single primer combination (dozen of such primer combinations can be used, one for each set of PCR reactions) detects upto 8 times more polymorphic bands and upto 16 times more loci, when compared with RFLP. More details about this technique and about its several modified forms are available in author's another text book. *"Biotechnology and Genomics"*. These markers have been used for construction of molecular maps in a variety of plant and animal systems.

Single Nucleotide Polymorphisms (SNPs) : Single nucleotide polymorphisms are due to single base pair variations or insertion/deletion of a few base pairs. The latter are sometimes also described as InDels. These are the latest of all the DNA-based molecular markers and are the markers of choice in the 21st century. Many methods are available for the discovery and detection of SNPs. These markers are also being used for construction of molecular maps. More details about these markers also are available in author's another text book *"Biotechnology and Genomics"*.

***Mapping Populations and Computer Softwares* :** The construction of a molecular map involves the following steps :

(i) selection of a molecular marker system;

(ii) developing a mapping population;

(iii) genotyping the mapping population with markers to be mapped (genotyping means, that for each marker, we will record data about the allele present in each individual of mapping population);

(iv) using the genotyping data for construction of map (with the help of a computer software). We have already discussed the different marker systems {step (i) above} and the technique for recording polymorphisms {step (iii) above}. The development of mapping population and the computer software that are used for construction of molecular maps will be discussed in this section, before we discuss the available maps for animal genomes.

Mapping Populations

The mapping population, which is used for construction of a molecular map may represent a panel of known families (as in humans) or may be developed through crosses between diverse parents, representing the same species or different related species. In humans, a panel of known families is used, since desired crosses can not be made and family size is too small to be adequate for mapping (see later for panel of families used in humans). In livestock and other animal systems, a mapping population is developed by making crosses to recover recombinants among as many markers as possible. The segregating mapping populations developed from intraspecific or interspecific crosses can belong to one of the following types:

(i) F_2 population derived from an intraspecific cross;

(ii) BC_1 (first generation back cross) population derived from an intraspecific cross;

(iii) BC_1 population derived from an interspecific cross (e.g. *Mus musculus* x *M. spretus* for a map in mouse) and;

(iv) a population of recombinant inbred lines (RILs) derived due to continuous sib-mating in segregating populations derived from an intraspecific cross; this population can be used just like segregating F_2 population, but is considered superior, because it results from several rounds of meiosis and is immortal.

Computer Softwares for Construction of Linkage Maps : Several computer programmes are currently available for the construction of multipoint linkage maps. These programmes are used extensively for a variety of animals and plants:

(i) In 1985, LINKAGE was used for preparation of several maps;

(ii) In 1987, MAPMAKER and CRI-MAP became available and were found to be more useful than LINKAGE and were used for mapping several chromosomes in humans, mouse and several plant materials. CRI-MAP takes advantage of more genotypic data, inferring genotypes, wherever possible. Regardless of computer programme that is used for map construction, LOD scores are used for quickly finding the best genetic order for large number of loci. (For LOD scores, see next para). It has been shown that the various computer programmes produce maps with comparable genetic orders.

For ordering and spacing of different RFLPs or other markers on the molecular map, models are fitted by maximizing the likelihood of certain parameters. The likelihood of two alternative models, one assuming presence of linkage, and the other assuming no linkage, are compared by calculating the likelihood ratios. When the likelihood ratio is 1000 : 1 (or at least 100 : 1) then the odds of one model are 1000 (or at least 100) fold greater than the odds of another, and therefore, the first model, assuming presence of linkage, is accepted.

The logio of likelihood ratio is called LOD score (e.g. for 1000 : 1 likelihood ratio, Logio = 3, LOD score = 3; for 100 : 1 likelihood ratio, logio = 2, LOD score = 2), and is used for linkage analysis. An LOD score of 1 to 2 is interesting, 2 to 3 is suggestive and > 3 is proof of linkage. It has been shown that no adjacent linked markers are separated by more than 32% recombination.

Genetic Maps using Molecular Markers : Genetic maps using molecular markers have been successfully prepared for all important animal systems including humans, mouse, rat and fruitfly. As we know, distances between markers in these maps are proportionate to recombination frequencies, and genetic distances for the map are worked out on the basis of a mapping function. Although initially, in early 1980s, RFLPs were the only molecular markers that were available for preparing molecular maps, later a variety of other PCR-based molecular markers including AFLPs, SSRs and SNPs were used extensively for developing these maps.

Molecular Genetic Maps in Humans (Homo Sapiens) : For the construction of the primary genetic linkage maps for humans, a panel of 59 families (27 from Utah, USA; 29 from France; 3 from Venezuela) containing 1212 informative meioses was used. To provide a permanent source of DNA, cell lines were established from these 59 families. Centre for Study of Human Polymorphism or CEPH (Centred'Etude du Polymorphism Humain) in Paris (France) had undertaken to distribute DNA from these cell lines to laboratories collaborating to map the human genome. (CEPH was created by Jean Dausset using money for his Nobel Prize and a large legacy). The decision to use the same material allowed markers studied in different laboratories to be combined in a single linkage map.

Primary genetic linkage maps with markers spaced approximately 15 cM apart were initially (1992) available for most human chromosomes. These primary genetic maps were then used for the construction of high resolution linkage maps with a variety of markers (e.g. RFLPs, SSRs, SNPs) spaced only 0.5-2 cM apart.

List of Some Molecular Maps in Humans

Group	*Year*	*Average spacing (CM)*	*Marker (no. & types)*
1. Collaborative research	1987	10	403 loci (including 393 RFLPs)
2. N1H/CEPH	1992	<5	146 loci (including 279 genes)
3. Genethon, Paris	1992	4.4	814 microsatellites
4. Pittsburgh	1994	6.2	655 loci
5. CHLC	1994	4.9	123 loci
6. Genethon	1994	2.9	2,066 microsatellites
7. HGP	1995	-0.2	15,000 STS
8. Genethon	1996	—	5.264 microsatellites
9. Genethon and others	1998	-0.5	8.031 STRPs*
10. IHGSC	1998	—	30,000 genes
IRHMC	1999	—	GeneMap, 99
TISMWG	2001	—	SNP map : 1.42×10^6 SNPs

* *STRPs = microsatellites = SSRs*

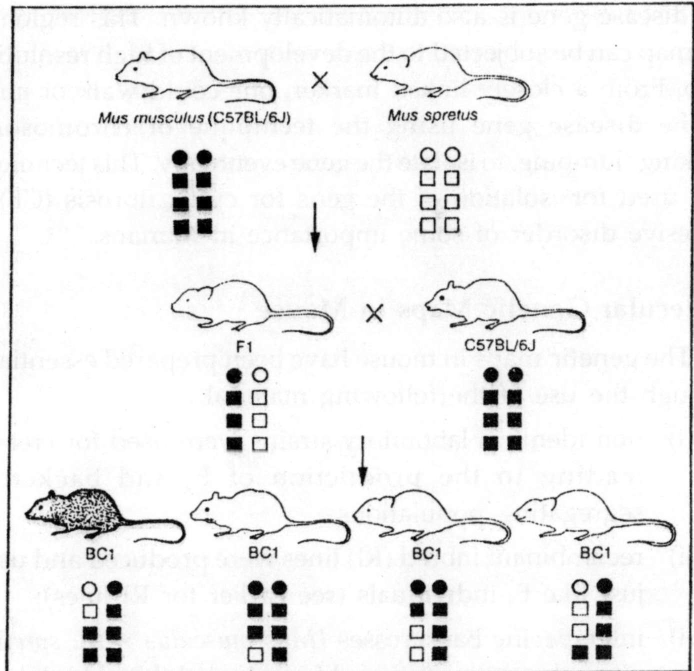

Breeding scheme used to generate the interspecific backcross mapping panel; for comparison only one pair of chromosomes is shown; M. spretus alleles are shown by open boxes, M. musculus (C57B/6J) alleles are shown by closed black boxes; note that the recombination events of F. can be scored in BC, backcross progeny.

Such high resolution maps have already been prepared for all human chromosomes (Table). The most comprehensive genetic map having 8031 short tandem repeat polymorphisms (STRPs) was prepared by Genethon, the University of Utah and the Cooperative Human Linkage Centre (Table). A much denser map of 1.42 million SNPs became available in 2001.

The primary linkage map can be utilized for locating genes for a specific disease on a specific region of a chromosome. This is done through the study of cosegregation of disease with a molecular marker in a reference family. Once linkage of disease with a marker is known through a study of cosegregation, if the marker is already mapped, the chromosome region carrying the disease gene is also automatically known. This region of the map can be subjected to the development of high resolution map. From a closely linked marker, one could walk or jump to the disease gene using the technique of chromosome walking/jumping, to isolate the gene eventually. This technique was used for isolation of the gene for cystic fibrosis (CF), a recessive disorder of some importance in humans.

Molecular Genetic Maps in Mouse

The genetic maps in mouse have been prepared essentially through the use of the following material :

(i) non-identical laboratory strains were used for crosses leading to the production of F_2 and backcross segregating populations;

(ii) recombinant inbred (RI) lines were produced and used just like F_2 individuals (see earlier for RI lines):

(iii) interspecific backcrosses *(Mus musculus* × *M. spretus)* were also utilized, since *M. spretus* exhibited high level of genetic divergence with *M. musculus.*

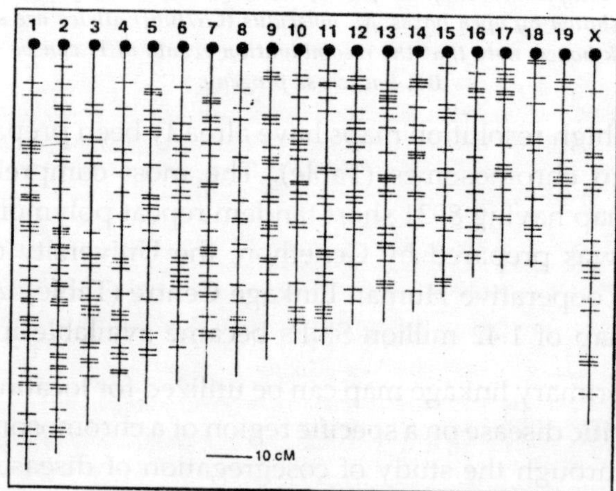

Diagrammatic representation of interspecific backcross linkage map of mouse; mouse chromosomes are acrocentrics (centromeres are indicated by filled circles); the distances between centromeres/telomeres and markers are not known.

The mouse has a genome, which is as big as that of humans. RFLPs, RAPDs and SSRs (microsatellites) were used as molecular markers. In 1989, a linkage map with 965 loci was presented. This map included phenotypic markers, biochemical markers, cloned genes, and anonymous DNA markers. Subsequently in 1991, 600 loci were positioned on the map using interspecific crosses, which gave a resolution of 3 cM. More than 300 SSR loci were also mapped in 1992. By 1996 a dense genetic map with nearly 6600 highly polymorphic SSLP (SSLP = SSR) became available. Subsequent efforts led to saturation of a map which had as many as 12,000 polymorphic markers at the end of the year 2002. Most of this work was done in USA, by MUGO (*Mus musculus* Genome Organization).

Molecular Genetic Maps in other Animal Species : Molecular genetic maps are also being prepared in other farm animals for commercial and medical reasons. These maps of farm animals will enable genetic engineering for breeding programmes to increase yield, growth rate and other commercially important traits. It has also been shown that microsatellites are conserved in related animals. For instance, 50% of cattle microsatellites are amplified by primers from sheep and *vice versa.* Therefore, ' heterologous primers are being used for development of SSR or microsatellite maps of related species. Mouse and rat models can also be used for elucidating the genetics and biochemistry of human diseases. Therefore, development of molecular genetic maps in these rodents will be a great help in human and livestock health care.

Cytogenetic Maps using Molecular Markers : The first step in mapping of chromosomes at the molecular level is to locate the markers on specific chromosomes. If linkage groups corresponding to individual chromosomes are available, then the molecular markers may be assigned to these chromosomes through study of linkage of available markers with molecular markers already mapped. In other cases, hybrid cell lines of known cytogenetic composition may be used. These hybrid

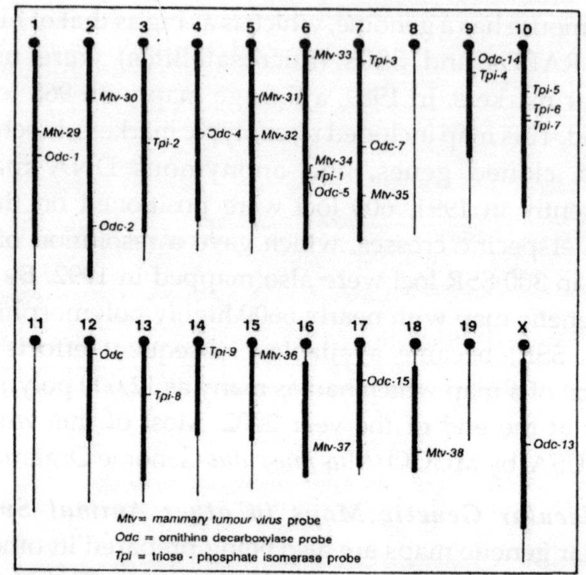

A mouse linkage map showing approximate locations of molecular probes Mtv (mammary Tumour virus), Ode (ornithine decarboxylase) and Tpi (triose-phosphate isomerase) loci on mouse chromosomes.

cell lines often have each, a full set of chromosomes from one rodent (like mouse) and only one known specific human chromosome, which thus enables us to assign a marker or a protein encoding gene to a specific human chromosome. This has allowed the construction of meiotic linkage maps for each of the 24 (22 autosomes + X + Y) human chromosomes. These maps have proved useful for locating genes for specific human diseases on specific linkage sites of individual chromosomes. In some cases, these maps also helped in the isolation of specific genes. Despite these advances, on the available maps, 1 cM is still equivalent to 1 megabase (Mb = 10^6 bp) and *in situ* hybridization can not resolve less than several million base pairs. Pulsed field gel electrophoresis (PFGE) is also being used but due to technical limitations, DNA sequences more than a few hundred kilobase pairs can not be ordered. These techniques and some of the results achieved through these techniques will be discussed in this section.

Cytogenetic Maps using Chromosome Specific Markers

Use of Somatic Cell Hybrids of Known Chromosome Constitution : Chromosome specific markers could be isolated from libraries prepared from somatic cell hybrids like mouse-human or hamster-human hybrids, which contain only one human chromosome of known identity (e.g. chromosome 10, 16, 20, 21, 22 etc.) The libraries prepared from such somatic cell hybrids were screened for those clones which contained human DNA inserts (using whole human genomic DNA as a probe), so that these selected clones will carry DNA sequences from a specific human chromosome. For instance, in an experiment, mouse-human hybrid cell line CHI-16, containing human chromosome 16 as the only human component, was used to construct a cosmid library with DNA fragments, each of approximately 40 kilobase pairs in length. From this library, cosmids containing chromosome 16 specific human DNA inserts could be selected through hybridization with ^{32}P-labelled human genomic DNA. These cosmid probes when used for mapping would give a cytogenetic map for chromosome 16 and so on.

Use of Flow Sorted and Microdissected Chromosomes : In mouse and human, individual chromosomes can be separated from mouse-human hybrid cell lines using the technique of flow sorting. Both normal and translocated human chromosomes could be obtained through flow sorting. In mouse, the chromosomes being similar in size, only five peaks were obtained in flow sorting and individual chromosomes could not be easily separated except when all chromosomes other than chromosomes 19 and X were fused to produce metacentric chromosomes. This technique allowed the construction of chromosome specific libraries in γ (lambda) vector. X-chromosome specific cDNA clones were also identified by probing γ -phage genomic libraries from normal humans with P labelled cDNA probes derived from hamster-human cell hybrid cell lines containing the X chromosome as the only human chromosome. Only X-specific cDNA clones will hybridize, which will represent the transcribed sequences on X-chromosome and will thus allow the construction of a map of X-chromosome, involving functional genes only. A phage library was also prepared using flow sorted chromosome material enriched for chromosome 16.

Micromanipulators have also been used for isolating specific chromosomes (where specific chromosomes can be identified), both in animals and plants. Once these specific individual chromosomes are separated in large number, they can be used for microcloning to give chromosome specific clones. Fragments of chromosomes, representing specific regions, can also be dissected physically from metaphase spreads of chromosomes and used for microcloning. This will give chromosome region specific (or band specific) libraries to be used eventually for a variety of purposes including the physical mapping. For instance, in *Drosophila*, polytene chromosomes have been used for microdissection and microcloning, leading to the preparation of physical maps. Techniques have also been developed for precise dissection of single bands from human chromosomes, using siliconized glass needles and electronically controlled micromanipulator. The DNA from these chromosome fragments is then extracted, digested with an enzyme, cloned, and amplified using polymerase chain reaction (PCR) to give specific DNA libraries. As few as 20 dissected chromosome fragments are enough to give such a library. Since the dissected chromosome fragments will not give enough quantity of DNA on extraction, PCR technology for amplification has particularly proved very useful in this connection. These chromosome sorting and microdissection techniques can be easily used for construction of chromosome specific or band specific genomic libraries, leading to the construction of chromosome maps.

Assigning Protein Coding Genes to Chromosomes : Assignment of protein encoding genes to specific chromosomes is a major goal of human cytogenetics. This can be routinely achieved by hybridizing a cloned gene with Southern blots or dot blots prepared from DNA samples extracted from hybrid cell lines of known cytogenetic composition. The Southern blots in this case allow rapid identification of a chromosome specific restriction fragment of the gene of interest. However, in this exercise, pseudogenes and genes related with the probe or sequences in a highly conserved gene may give a false positive signal, so that not all restriction fragments so identified may represent the gene of interest. Therefore, a gene may not be assigned to a specific chromosome with complete certainty. In view of these difficulties in chromosome mapping, an alternative approach was utilized to map the gene P68 in humans. For this gene, the conserved region was already known and was used for the synthesis of primers to be used in PCR amplification. Speciès specific differences between mouse and humans could be detected in PCR products, the difference (15bp difference) being due to intron lengths. This property was used in screening a number of hybrid cell lines of known cytogenetic composition to find out the chromosome on which the gene may be located. The hybrid cell line PCTBAI.8 contained only one human chromosome i.e. human chromosome 17 and DNA from this line gave a human specific restriction fragment in PCR product thus confirming that *p6S* gene was present on chromosome 17. Later, hybrid cell lines having either only one or more fragments of human chromosome 17 were derived. The use of these cell lines enabled the mapping of *p68* gene to the distal part of the long arm of chromosome 17 in the region between the bands 17q23 and 17q25.

Physical Maps using Molecular Markers : The physical maps of DNA fragments, chromosome fragments or of whole individual chromosomes are based on distances between genes or distances of genes from centromeres or from specific molecular markers measured either in terms of base pairs (megabase pairs = 10^6 bp) or in terms of relative physical lengths. For instance, the distance of a gene from centromere may be expressed as a fraction of the length of the whole arm taken as unity, so that the distance of a gene from centromere will be indicated as 0.1, if it is located at a distance of $1/10^{th}$ of the whole arm. Sometimes, physical maps of individual loci on small chromosome fragments were also prepared, where the DNA sequences on either side of the locus are shown in the form of restriction sites for a number of restriction enzymes. This could be extended in some cases to the preparation of restriction map of a whole chromosome. Therefore, one should realize that in physical maps, distances are never expressed in centiMorgan units which are proportionate to recombination frequencies. The distances in physical maps expressed in kilobase or megabase pairs are sometimes not proportionate to recombination frequencies and, therefore, the physical maps are considered more important than the genetic or linkage maps. Some of the techniques used for preparation of these physical maps will be described in the following sections.

Top Down (Restriction Mapping) vs Bottom Up (Contig Mapping) Approaches : One of the early goals of what was later called *'Human Genome Project'* was to make physical maps of each of the 22 autosomes and two sex chromosomes (X, Y) in humans. Two conceptually different types of strategies were tried : (i) The first strategy involved preparation of restriction maps of large DNA fragments (macro restriction analysis), generated by rare enzymes that cut the genome infrequently. It is achieved through analysis of DNA fragments by pulse-field gel electrophoresis (PFGE). This approach is reductionist or top down approach, because we start with isolation of individual chromosomes or their large segments using hybrid cell lines of known chromosome composition. Then from the existing genetic maps of these isolated individual chromosomes (or segments), we use probes that serve as anchor points on the physical map. The gaps between these anchor points are then filled by extensive Southern analysis. This technique initially yielded restriction maps for yeast *(Schizosaccharomyces pombe)* chromosomes; (ii) The second strategy involved preparation of complete ordered genomic library. This is a bottom up approach, because we start with individual clones, which are fingerprinted and through detection of overlapping of DNA segments in clones, the latter are arranged in a large number of contiguous sets of overlapping clones (called contigs), which coalesce into a map of the whole genome. This bottom up approach has been used for whole genome sequencing (see later for details).

Above two approaches are complementary, but have their limitations and benefits. For instance, the *top down* approach of restriction mapping of large DNA fragments, provides information about distantly spaced genomic regions. However, it does not provide purified DNA in an immortal amplifiable form. In the second approach *i.e. bottom up* approach of ordered complete library, on the other hand, it is easy to start with individual clones, but it is difficult to complete the library, there being many gaps. However, it provides a high resolution information and a permanent source of purified DNA. It is believed that both types of maps are useful and have been constructed in parallel.

Physical Maps using Chromosomal Aberrations (deletions): Physical mapping of specific DNA sequences available in the form of probes, could be achieved using chromosomal aberrations such as translocations, duplications and deletions. For instance, the disease choroidermia or tapetochoroidal dystrophy (TCD) was described in various patients having cytogenetically visible or submicroscopic deletions spanning the X_q21 band. This meant that TCD locus must fall within the region of these overlapping deletions. In order to determine the location of TCD locus more precisely, several of these deletions were characterized with the help of several anonymous cloned DNA sequences from the proximal region of X_q. This led to the division of X_q21 band into seven different intervals and TCD was assigned to interval 3 spanning three probes (DXS96, DXS165, DXS233). It was also found that several TCD patients had deletions which included the locus DXS165 but not the other two loci. This suggested that TCD may be close to DXS165. Further mapping was achieved by chromosome jumping and walking and TCD was assigned to a DNA segment of 15 to 20 kilobases. Since, DNA fragments, several hundred kilobase pairs long, could be cloned in YAC vectors, a YAC library of 1500 clones (instead of 10,000 with cosmids) was sufficient for mapping the *Drosophila* genome (180Mb). During the years 1989-92, as many as more than 1000 *Drosophila* YAC clones (called DY clones) were cytologeally locate through *in situ* hybridization (ISH) of these clones on salivary gland chromosomes. Physical maps using YACs were also prepared in humans and mouse.

Physical Maps using in situ Hybridization with Yeast Artificial Chromosomes (YACs) : Development of YACs as vectors for large DNA fragments. Using these YACs, molecular maps in human, mouse and *Drosophila* were prepared. Actually, for large scale physical mapping of eukaryotic genome, *Drosophila* is considered a model organism (though not as attractive as yeast or *Aspergillus).* Since, DNA fragments, several hundred kilobase pairs long, could be cloned in YAC vectors, a YAC library of 1500 clones (instead of 10,000 with cosmids) was sufficient for mapping the *Drosophila* genome (180 Mb). During the years 1989-92, as many as more than 1000 *Drosophila* YAC clones (called DY clones) were cytologeally located through *in situ* hybridization (ISH) of these clones on salivary gland chromosomes. Physical maps using YACs were also prepared in humans and mouse.

Physical Maps using Chromosome Walking and Jumping (or hopping) : The centromeric sequences on a yeast chromosome could be isolated using chromosome walk from two cloned genes *CDC10* and *LEV2* on either side of the centromere. Walks have also been used to establish the structure of some very large *Drosophila* genes that are much too long to be cloned even in cosmids. Unfortunately, chromosome walks are difficult in higher eukaryotes, due to the presence of repetitive sequences flanking majority of the genes, which allow cross hybridization. Therefore, initially the fragments of DNA could not be arranged along the length of chromosomes in higher eukaryotes.

Chromosome jumping technique was also used to jump large distances to approach closer to a gene of interest. Size selected DNA fragments were generated by partial *MboI* digestion and a jumping library was prepared. Jumping has the additional advantage of allowing one to jump over otherwise unclonable sequences.

In humans, following jumping libraries, besides others, were prepared, (i) 100 kb jumping library to move towards *Cystic Fibrosis* (CF) locus, (ii) 200 kb jumping library to move toward *Huntington Disease* (HD).

In these cases, clones were derived which jumped from a given locus (probe) towards the gene of interest. The jumping library (with 2 10^6 clones) was screened using closely linked probes that were labelled. Positive phage clones were subcloned and used for further chromosome jumping.

The gene responsible for *Huntington Disease (HD)* in humans, is one such gene that was identified through its close linkage to the DNA marker *DHS10*, which by *in situ* hybridization and somatic cell genetics, was mapped near the tip of the short arm of chromosome 4 (4p). Chromosome jumping was used to move from *DHS10* to *HD*. Similarly, cystic filorosis gene was also mapped using chromosome jumping.

Physical Mapping of Chromosome Ends using Telomeric DNA Fragments : A universal deficiency in genetic maps of higher organisms is the absence of markers that identify the ends of chromosomes. Unique, physically defined genetic markers associated with individual telomeres would provide the boundaries for genetic maps. In view of this, availability of a molecular marker representing the telomeric sequence, will be a great help in preparing genetic and physical maps in eukaryotes. This telomeric sequence has been found to be a six nucleotide sequence that is repeated in the telomeric region over and over again without mistake. The sequence has been isolated in *Arabidopsis, Homo sapiens, Tetrahymena, Paramecium, Oxytricha, Saccharomyces,* etc. and is G-and-C rich with a general formula (T or A)$_m$ (G)$_n$, which is (TTAGGG)$_n$ in humans. When this telomeric sequence was used for *in situ* hybridization, the hybridization took place in a wide variety of vertebrates, showing that the sequence is conserved and variation though exists, but should be only minor. Chromosome specific subtelomeric sequences for human chromosomes have also been cloned, which can be used to find the ends of specific chromosomes.

STS-based Physical Map of Human Genome : The US-Human Genome Project, has set a target of a physical map consisting of 30,000 STSs (sequence tagged sites), spaced at interval of 100 kbp. Towards this end, in the year 1995, a physical map of human genome containing more than 15,000 STSs was published, which had an average spacing of 199 kilobases. These STSs were developed from radiation hybrid map and linkage map of human genome.

BAC Clone Physical Maps in Humans : Physical maps of human genomes in the past were prepared for only parts of the genome or for individual chromosomes at different levels. For instance, in 1998, a physical map of 30,000 human genes was prepared, and in 1999, a new human gene map (based on radiation hybrids) called GeneMap'99 was prepared by The

International Radiation Hybrid Mapping Consortium (IRHMC). However, IHGSC (International Human Genome Sequencing Consortium) centres also prepared the whole-genome clone-based physical maps for human genome, which facilitated sequencing of the human genome (see later in this chapter, for the need of BAC physical map for whole genome sequencing vs whole genome shotgun sequencing, which does not need any map). In early efforts, clone based regional and chromosomal physical maps were prepared, with limited success. However, physical maps based on sequence-tag site (STS) landmarks provided greater coverage of the genome. Later, the use of PAC and BAC clones, having larger inserts and better stability, proved helpful in achieving success in the preparation of a whole genome physical maps (for BAC and PAC vectors).

Whole-genome BAC Map : Initially (during 1995-98), physical maps were prepared for specific regions of the genome. STS markers were used from these regions to screen PAC and BAC clones and the positive clones were fingerprinted for mapping. However, subsequently whole genome physical maps were prepared using several steps, which included the following:

(i) fingerprints of a large number of random clones from a genome-wide BAC library were collected;
(ii) fingerprint data was used to cluster highly related clones, which were then assembled into contigs;
(iii) the contigs were merged with related clones at their ends;
(iv) the individual clones and the corresponding contigs were mapped using landmarks from the existing maps.

Fingerprinting the BAC Clones and Assembly of Map : From a BAC library (RPC-11), as many as 300,000 BACs having an average insert size of 150,000 bp were fingerprinted during 1998-99. This gave a 15-fold coverage of the genome ($3 \times 10^9 \times 15 = 45 \times 10^9$; human genome = 3×10^9 bp). Using high throughput approach for fingerprinting, as many as 20,000 fingerprints were obtained every week, so that by the end of 1999, the data on fingerprints of 300,000 BACs were available in FPC (htttO://www.sanger.ac.uk/software/lpc/). Eventually, using strategies for automated assembly, 283, 287 BAC clones were assembled in ~7,700 clusters and the remaining fingerprints were excluded. The clones within clusters were then ordered into contigs, using overlapping, regions. Further refinement allowed drastic reduction in the number of contigs from 7,700 to 965, which were distributed along all the chromosomes.

Integration with other Maps : As many as 13,695 selected markers from different available maps (CEPH Genethon genetic map, GeneMap'99 = radiation hybrid map, etc.) were hybridized against colony filter replicas of the RPCI-11 BAC library. This permitted positioning of 96,283 BACs as genome anchor points for the contigs. As many as 9,018 STSs derived from BAC end sequences were also assigned to chromosomes. Integration of maps was also achieved by mapping 7,600 large insert clones (mainly BACs) by fluorescence *in situ* hybridization (FISH). These were used as cytogenetically defined landmarks on the draft sequence of the human genome (Nature 409 : 953-958 2001). As the sequence data accumulated, markers could also be identified using electronic PCR (ePCR), which searches for STSs by the associated primer sequences and verifies the presence of specific STSs. Eventually, the contigs were oriented with respect to majority consensus of all available maps including the following six maps : (i) Genethon genetic map (1996); (ii) Marshfield genetic map (1998); (iii) WIBR YAC STS-content map (http://carbon.wi.mit.edu:8000/Cgi-bin/contig/phy-map); (iv) Gene-Map'99; (v) SHGC G3 radiation hybrid map (http://www.shgc.stanford.edu/ Mapping/rh/index.html) and (vi) NCBI framework map (http: //www.ncbi.nlm.nih.gov/genome/guide/). Data available for the regional maps for parts of different chromosomes were also integrated in this map. Computer-generated restriction digests or *in silico* digests derived from sequences in GenBank were also used to incorporate mapped clones into the whole genome BAC map.

Use of physical map for selecting BACs for whole genome sequencing. For whole genome sequencing, the whole genome BAC physical map has been used in the past and continues to be used at the end of the year 2001. The map allows selection of BACs for sequencing using the principle of minimal tiling path.

Sequence Maps : SNP Maps in Primates : Earlier in this chapter, we briefly described SNPs (single nucleotide polymorphisms) as the third generation molecular markers, that became popular at the beginning of the 21st century. In the year 2001, along with the publication of the draft sequences of the human genome, a map of 1.42 million (1,420,000) SNPs, providing an average density of one SNP every 1.9 kilobases was also published (Nature, 15 Feb. p. 928). The map was prepared by 'The International SNP Map Working Group' (TISMWG) and it contained all SNPs that were publicly available in November, 2000. Similarly, an SNP map for mouse genome (with -2000 SNPs) was published in the year 2000, and a rat genome SNP map (with more than 12,000 gene-based SNPs derived from cDNAs) was published in the year 2004.

Discovery of SNPs : Over 95% of the 1.42 million SNPs mentioned above were discovered by The SNP Consortium (TSC) and the public Human Genome Project (HGP). The remaining 5% were discovered by other centres. TSC discovered SNPs, using a publicly available panel of 24 ethnically diverse individuals. SNPs were identified as follows:

(i) TSC identified 1,023,950 candidate SNPs from the sequences of genomic fragments, obtained by shotgun sequencing; these sequences were aligned to one another and to the available genome sequences, followed by detection of single base differences, using validated algorithms (polybayes and neighbourhood quality standard or NQS);

(ii) HGP identified 971,077 candidate SNPs from sequence differences in regions of overlap between large-insert BAC or PAC clones, that were sequenced.

Mapping of SNPs : SNPs were mapped, by aligning the sequences flanking these SNPs to the genome sequences of large-insert clones in GenBank (database). Since these genome sequences were already assigned to chromosomes, and because physical contig maps (for BACs and PACs) of chromosomes were available, SNPs could be assigned positions on chromosomes. Integrated maps of SNPs, genes and other features are available on websites as follows : ENSEMBL (http.// www.ensembl.org), NCBI (http./Avww.ncbi. nih.gov) and TSC (http://snp.cshl.org).

RefSeq database released in September 2000 was also used to find out the distribution of SNPs in transcribed and non-transcribed regions of the genome. RefSeq contained 7000 annotated non-redundant mRNA and 14,534 SNPs were mapped in these 7000 mRNAs giving a frequency of two exonic SNPs per gene, so that if total genes in human genome are 30,000, there should be 60,000 SNPs in the exonic regions. This gives a density of one SNP per 1.08 kb, which is higher than the density of SNPs in the genome as a whole (one SNP every 1.91 SNP).

Gene-based Sequence-tagged Sites (STSs) and Regional Gene Mapping : Gene-based sites have often been identified by using mRNA from a specific tissue and synthesizing complementary DNA (cDNA) from it. This cDNA is used to prepare cDNA library in the form of a number of clones. The 3' untranslated regions (3' UT) of these clones are sequenced and used not only to identify human genes, but also to generate gene-based sequence tagged sites (STSs). These STSs facilitate large scale PCR mapping of human genes. This approach was used for regional gene mapping on a genome-wide scale.

Novel High Density Maps using Radiation Hybrid (RH) and HAPPY Mapping : Due to technical limitations, DNA sequences more than a few hundred kilobase pairs long can not be easily arranged in a linear order. Therefore, high resolution maps can not be easily prepared. Radiation hybrid (RH) mapping is a method involving use of somatic cell genetics, which allows ordering of DNA markers having millions of base pairs of DNA, with a resolution of 500kb. RH mapping with PFGE (pulse field gel electrophoresis) has actually been used to construct a high resolution map of a 20 Mb (20×10^6 bp) DNA segment belonging to the long arm of human chromosome 21 (Cox et al., 1991). An *in vitro* technique analogous to RH mapping was HAPPY mapping, which was developed in 1993 and used for mapping in humans and some microbes. (For details about RH mapping and HAPPY mapping, readers should consult author's book *"Biotechnology and Genomics"*.

Integrated Genomic Maps

During 1980-2000, a number of molecular genetic, cytogenetic and physical maps were published for a number of animal systems, including human, mouse and fruitfly. Separate maps were generally prepared using different classes of molecular markers (e.g. RFLP, RAPD, SSRs, etc.), and more than one maps were published using the same marker system (e.g. microsatellites). The linkage groups were often assigned to chromosomes, thus generating the cytogenetic maps, and physical maps were prepared using a variety of approaches discussed earlier in this chapter. From time to time, the data from these multiple maps for the same genome (e.g. human, mouse, etc.) were also used for constructing integrated composite maps. These maps would have information on linkage distances (in cM), chromosome location, and physical distances in terms of number of base pairs. These maps are also being integrated with draft sequences of the genomes of human, mouse, rat and fruitfly, which became available during 2001-2004.

Linkage Disequilibrium (LD) Maps of the Future : At the beginning of the 21st century, mapping efforts are now focussing on LD mapping, where maps are not based on recombination frequencies, but on association of markers (for details of LD, consult a book on population genetics). The integrated maps are also being extended now to represent the extent of LD in different regions of the genome. LD mapping gained emphasis due to availability of whole genome sequences and SNP maps. These LD maps will prove extremely useful in the field of biomedicine including diagnostics, both in humans and livestock.

Whole Genome Sequences

During 1990s, high throughout DNA sequencing became possible, due to the availability of a variety of slab-gel and capillary based automatic sequencers. Consequently, methods for sequencing of large genomes of multicellular organisms were developed and several such genomes were fully sequenced. In this section, we first briefly describe the methods of whole genome sequencing and then describe the present status (as in the middle of 2004) of whole genome sequences of humans and several other animal systems. Since this area of research is fast expanding, the readers may need to update the information given in this section using the internet facility.

Methods for Whole Genome Sequencing : Methods for rapid and efficient sequencing of DNA fragments were developed in 1970s, which were described. The chain termination DNA synthesis involving ddNTPs (dideoxynucleoside triphosphates) developed by Frederick

Sanger has been preferred over other methods. Automated sequencers have also been developed to accelerate the rate at which sequencing data can now be generated. More recently new automated sequencers have been introduced, which make use of capillary separation rather than polyacrylamide gel separation of small synthesized DNA fragments. These new generation automated sequencers (e.g, ABI3700; MegaBACE 4000), each has 96 or even upto 384 channels thus permitting determination of 96 or 384 sequences in parallel. Since each run takes less than 2 hours, 1000 to 4000 sequences can be obtained from a single sequencing machine in one day. This method is now routinely used for shotgun sequencing of BAC clones, as discussed in the next section.

Whole-genome shotgun (WGS) sequencing. Whole-genome shotgun sequencing involves shearing or cleavage (partial digestion) of genomic DNA followed by cloning, to produce a genomic library. This is followed by sequencing of cloned DNA-fragments at random, followed by assembly of the fragment sequences into larger units on the basis of their overlaps. The technique is described as shotgun assembly and is illustrated in Fig.

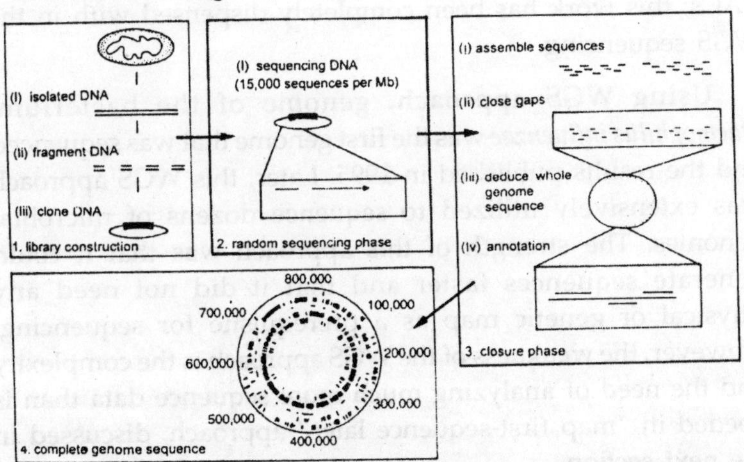

Different steps involved in whole genome sequencing (WGS) technique used for sequencing the genome of Haemophilus influenzae.

This approach does not require physical maps of the genome for whole genome sequencing. In the past, this approach was used for small genomes, each having very little or no repetitive sequences (e.g., viruses, bacteria, archaea, fruitfly, etc.), but later (during 1998-2001), this approach was extensively utilized by Craig Venter at a private company named Celera Genomics in USA (Rockville, Maryland) not only for sequencing *Drosophila* genome, but also for sequencing the human genome (Craig Venter in 2004, became the President of a new company, named "The Centre for the Advancement of Genomics" = "TCAG").

However, while using WGS sequencing approach for human genome sequencing, Craig Venter also made use of publicly available hierarchical shotgun DNA sequence data generated by the International Human Genome Sequencing Consortium (IHGSC). In this WGS approach, 8-10 fold sequencing of the whole genome is needed for coverage of the whole genome. Much less amount of sequencing (4-5 fold) is needed in hierarchical shotgun approach (consult next section), although in this latter approach, considerable work is involved in preparing the whole genome physical map using BACs and PACs; this work has been completely dispensed with in the WGS sequencing.

Using WGS approach, genome of the bacterium, *Haemophilus influenzae* was the first genome that was sequenced and the results published in 1995. Later, this WGS approach was extensively utilized to sequence dozens of microbial genomes. The strength of this approach was that it could generate sequences faster and that it did not need any physical or genetic map as a prerequisite for sequencing. However, the weakness of the WGS approach is the complexity and the need of analyzing much more sequence data than is needed in "map first-sequence later" approach, discussed in the next section.

***'Map First Sequence Later' or 'Clone by Clone' Strategy (Also Called "Hierarchical Shotgun Sequencing")* :** When the whole genome sequencing work on humans and other organisms was initiated in late 1980s, it was decided that large segments (clones) of genomic DNA (produced by partial digestion) may first be aligned in a linear order on the chromosomes as overlapping segments, which can then be used as landmarks for sequencing data.

The sequences of individual clones can thus be conveniently coalesced to obtain the DNA sequence covering an entire chromosome. Large DNA segments are cloned in BAC vectors and these BACs are used for construction of physical maps. Such a map is described as a physical map of the genome, since the physical position of each clone on a chromosome is defined in the form of ordered BACs, unlike the genetic maps where markers are placed on a linkage map or a chromosome map on the basis or recombination frequencies rather than on the basis of physical distance.

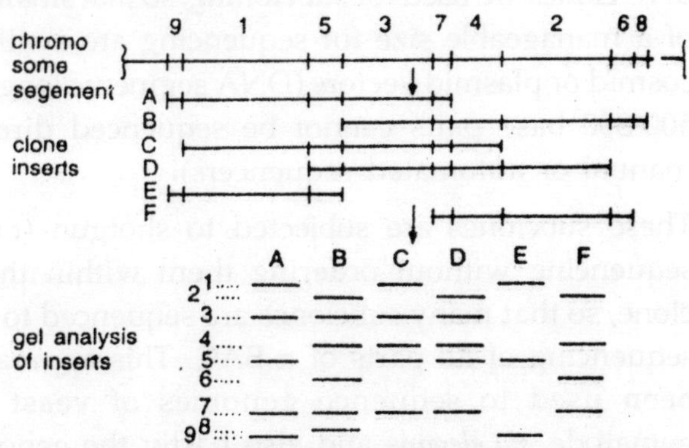

Construction of a physical map for BAC clones to be used for 'clone by clone' strategy of whole genome sequencing.

In late 1980s and early 1990s, such clone-based maps were considered necessary and useful for complete genome sequencing and were therefore prepared in several animal and plant genomes. Using these clone-based maps, whole genome sequencing was successfully completed in several eukaryotes including yeast *(S. cerevisiae)*, a nematode *(C. elegans)* and higher plant *(Arabidopsis thaliand)*. Such clone-based maps also contributed, though partly, to the whole genome sequencing of *Drosophila melanogaster*, the mouse and the humans.

Once the BACs are physically mapped, the physical maps can be utilized for whole genome sequencing using the following steps :

(i) BAC clones are selected from the whole genome BAC map, using suitable algorithms (software), so that minimum number of BAC clones with minimum overlapping is used to cover the entire genome. This is often described as selection of minimum tiling path. In case of human genome, 10,000 to 20,000 BACs were selected to generate a working draft of human genome;

(ii) BAC clones are used for subcloning, so that small inserts of a manageable size for sequencing are available in cosmid or plasmid vectors (DNA segments longer than 500-800 base pairs cannot be sequenced directly in manual or automated sequencers).

These subclones are subjected to shotgun (random) sequencing without ordering them within the BAC clone, so that many subclones are sequenced to ensure sequencing of all parts of a BAC. This approach has been used to sequence genomes of yeast and a nematode, *C. elegans* and also partly the genomes of fruitfly, mouse and humans. In this approach, every part of the genome is actually sequenced roughly 4-5 times to ensure that no part of the genome is left out (compare with WGS sequencing, in which 8-10 fold sequencing of the entire genome is needed).

Progress Towards Whole Genome Sequencing : Considerable progress has been made in the area of whole genome sequencing during the last decade of the 20th century and in the early years of the 21st century (Tables). In particular, considerable progress has been made in the study of genomes of animal systems including human beings (Fig). Among the genomes of multicellular organisms (Fig.) sequenced by 2004, the first that was sequenced in 1998 belonged to a nematode *(Coenorhabditis elegans)* and the second that was completely sequenced in February, 2000, belonged to fruitfly *(Drosophila melanogaster)*.

Later in February 2001, working drafts of human genome sequences were published; in spring 2001, and in December, 2002, drafts of the sequence of mouse genome became available.

Progress Towards Complete Genome Sequencing of some Important Organisms (as in July, 2004)

Organism	*Genome size (Mb)*		*% genome sequenced*
E. coli	4.8		100
S. cerevisiae	14.4		100
C. elegans	100.0		100
D. melanogaster	120.01	(euchromatic part only)	100
M. muscutus	2500.0		drafts
H. sapiens	2900.0		drafts
R. norvegicus	2750		drafts
A. thaliana	130.0		98%
O. saliva (*Japonica and indica*)	430.0		drafts

In the year 2002, the genome of house fly *(Anopheles)* was also sequenced. Complete refined sequences (finished and gap free sequences) of human genome are expected by the year

2004 and those of mouse genome are expected by the year 2005. Sequencing of rat genome was also started by three laboratories (one academic centre and two private companies) in USA with the sanction of $60 million by the US government for this work. The first draft of rat genome sequence was published in 2004. American scientists also completed a draft of the chimpanzee genome and made it available to researchers worldwide via the internet in early 2004. This sequence would be published by the end of 2004. It is believed that the whole genome sequences of these important mammals in hand will greatly facilitate biomedical research by unravelling the genetics and physiology of human diseases. Genomes of several other multicellular animals are also being sequenced. The information generated from these genome projects will be briefly presented in this section.

C. elegans genome. *Coenorhabditis elegans* is a nematode worm that has a small genome of 100 Mb (100×10^6 bases). During mid-1960s, the worm was selected as a model system for the study of nervous system, because it was perhaps the smallest multicellular animal, the adult having only 959 cells (including about 300 neurons). In the year 1998, the genome of this worm was the first multicellular animal genome to be completely sequenced, after 8 years of effort, by two groups, one located at Washington University Genome Sequencing Centre in St Louis, Missouri (USA) and the other located at Sanger Centre in Cambridge (UK).

The genome was sequenced with the help of a genomic library. The clones from this library were fingerprinted and overlapping regions of the clones were identified using landmarks on these clones. This enabled the preparation of a physical map of these DNA fragments, so that 80% of the genome was sequenced utilizing these clones that could be multiplied in bacteria. The remaining 20% of the genome was sequenced using YACs (yeast artificial chromosomes) that are multiplied in yeast cells.

Data on Sequenced Eukaryotic Genomes (modified from Nature, February 15, 2001).

Organism	*Year*	*Genome size (Mb)*	*Bases sequenced*	*Total coverage (%)*	*Euchromatin covered (%)*	*Number of genes (predicted)*	*Number of genes (per Mb)*
Saccharomyces cerevisiae	1996	12	12	93	100	5,800	483
Coenorhabditis elegans	1998	97	97	94	100	19,099	197
Drosophila melanogaster	2000	180	116	64	97	13,601	117
Arabidopsis thaliana	2000	125	115	92	100	25,498	221
Homo sapiens							
Rough draft (HGP)	2001	2900	2,693	84	90	31,780	12
Rough draft (Celera)	2001	2900	2,654	83	88-93	39,114	15
Mus musculus : high quality draft (MGSC)	2002	2500	2370	95	90	22,011	9
Ratlus norvegicus	2004	2750	2600	90	90	20,973	8

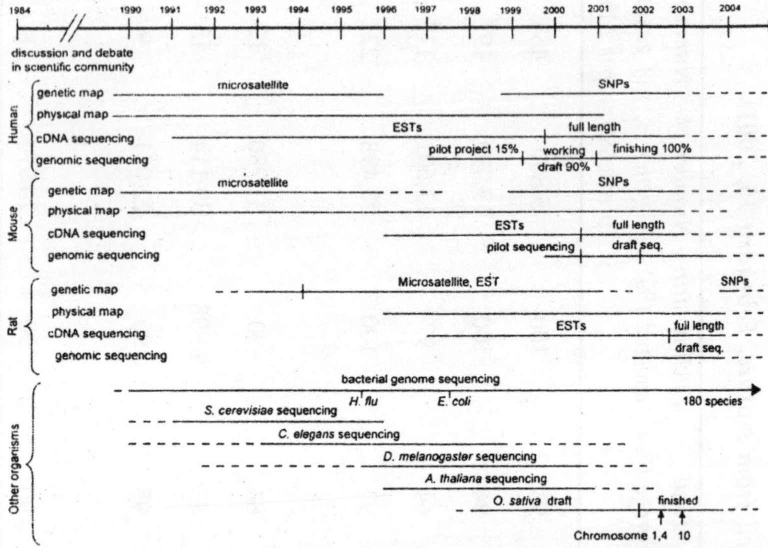

Progress of large scale genome analyses involving several non-vertebrate model organisms, the mouse, the human and the rat during 1990-2004.

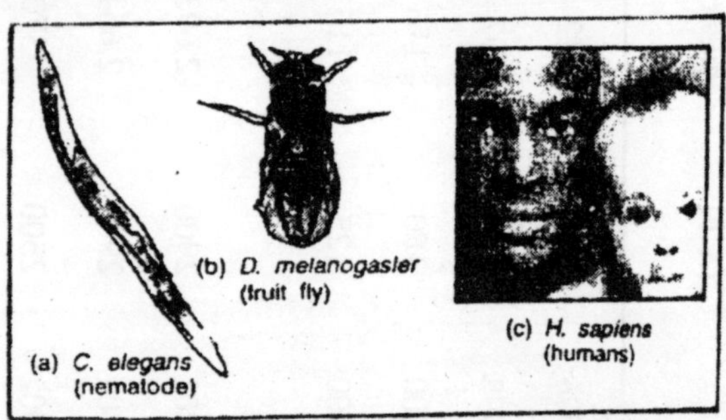

Three animals, whose genomes were fully sequenced by the year 2001 (in case of human genome, only draft sequences were released in the year 2001).

The genomic sequences of *C. elegans* have been compared with those of yeast and other microbes, revealing similarities and differences. The number of genes in the worm genome turned out to be approximately 19,000, three times those in yeast genome (6,200 genes). Although, the functions of 12,000 of these 19,000 genes were still a mystery in 1998, when the sequencing of the genome was completed, the information about these genes is being utilized by neurobiologists, biochemists, developmental biologists and evolutionary biologists. In order to find out functions of the above 12,000 genes (from a total of 19,000 genes) with unknown functions, a large number of mutant worms (-500,000) were produced at the University of British Columbia, Vancouver (Canada). These worms were studied for the defects, they had. These mutants along with RNAi (RNA interference) technology and GFP (green fluorescence protein) were used to find out how do individual genes controlling specific traits function. Microarrays for all the 19,000 genes have also been prepared to study the functions of not only all these genes, but also the sequences that regulate the activities of these genes.

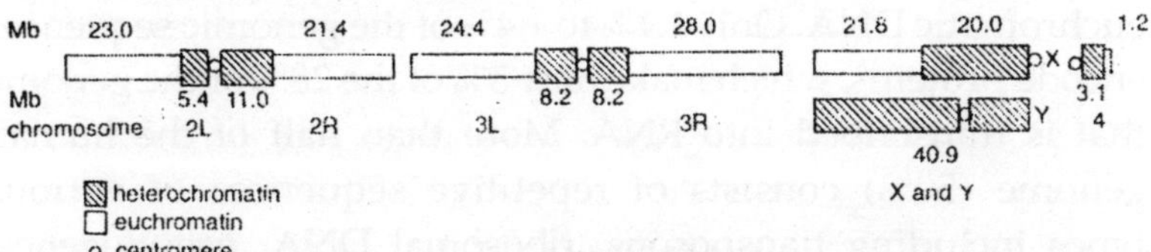

Mitotic chromosomes of Drosophila melanogaster showing euchromatic regions, heterochromatic regions and centromeres; 2L, 2R, 3L, 3R and 4 are chromosome arms; X and Y are sex chromosomes with heterochromatic blocks.

***Drosophila* Genome** : *Drosophila* genome is -180 Mb (Mb = 1 × 10^6 bp) in size, one third (60 Mb) of which is centric heterochromatin. The remaining 120 Mb of euchromatin is present mainly in two large autosomes and the X-chromosome; the small chromosome IV contains a meagre 1 Mb of euchromatin. Therefore, in *Drosophila,* it was 97% of only the 120 Mb euchromatic part of the genome that was sequenced jointly by three groups including the following :

(i) Berkeley *Drosophila* Genome Project (BDGP),

(ii) European *Drosophila* Genome Project (EDGP) and

(iii) the private company Celera Genomics.

The next step was to find out the functions of all *Drosophila* genes that constitute 120 Mb of genomic DNA that was sequenced. In the jargon of genomics, this process is described as 'annotation'. This was done by a team of workers including computer experts and *Drosophila* geneticists, who were described as annotation 'jamboree'. The annotation work was also completed quickly and it was shown that there were perhaps -13,600 (this figure was later revised to 13,061) genes in the euchromatin portion of *Drosophila* genome.

Human *(Homo sapiens)* genome. The human genome is 25 times the size of the eukaryotic portion (120 Mb) of *Drosophila* genome (180 Mb), whose successful sequencing was discussed in the previous section. Precisely its size is 3.2 Gb (1 Gb = 10^9 bp) spread over 24 chromosomes including 22 autosomes, an X-chromosome and a Y-chromosome. It contains 2.9 Gb of euchromatic DNA. Only 1.1% to 1.4% of the genomic sequences encode proteins, which make only 5% of the 28% of the genome that is transcribed into RNA. More than half of the human genome (50%) consists of repetitive sequences of various types including tiansposons, ribosomal DNA, pseudogenes, SSRs, centromeric repeats and telomeric repeats. Majority of these repeats consist of different classes of parasitic DNA elements {transposons and retroelements (LINEs, SINEs, LTR elements)}.

During 1986-87, there was a debate among molecular biologists about the need and desirability of working out the nucleotide sequence of the entire human genome consisting of -3200 Mb. It was also estimated that this will involve 32,000 person years of work at a cost of 2 billion US dollars. Despite this magnitude, the project was initiated on October 1, 1990 as an international effort with funds made available by several countries. National Human Genome Research Institute (NHGRI) in USA made major effort in this direction. The following works were undertaken under this project :

(i) Construction of a comprehensive genetic map of human genome;

(ii) Construction of landmarks based physical maps of human genome;

(iii) Construction of a human transcript map with positions of about 30,000 genes. These efforts also led to the identification of at least 200 disease associated genes. This information is of direct relevance to human welfare and will help in (i) understanding the diseases at the molecular level, and in (ii) the development of preventive strategies, treatments and cures.

Originally, the sequencing of the whole human genome was targeted to be completed by the year 2005, but the situation dramatically changed in 1998 with the creation of a private biotech company Celera Genomics by Craig Venter (earlier working in the public funded Human Genome Project), who announced in May, 1998 that he would complete sequencing of human genome by using WGS approach within three years, i.e. by the end of the year 2001. This created competition between public and private efforts, more so because if the private company had done it earlier, they would have proprietary rights over the whole genome sequence thus depriving the public organizations from utilizing, free of cost, these sequences for public good. In view of the competition, public Human Genome Project also redefined their targets and announced that they will complete the sequencing of the whole genome by the year 2003 instead of 2005, but it was still two years behind the deadline earlier announced by the private company Celera Genomics.

Fortunately, the dialogues between public and private projects led to simultaneous announcement of the release of first drafts of the human genome on June 26, 2000. Later, the two drafts, one each by the public and private efforts, were published simultaneously on 15/16 February, 2001 in the two most prestigious science journals of the world (Nature and Science).

Developments (in a Chronological order), Leading to Discovery of Complete Sequence of Human Genome.

Year	*Scientist/organization*	*Development/activity*
1985	Charles Delisi, Associate Director, Health & Environment Research, DoE, USA	Initial discussion to sequence the complete human genome; DoE funding started in 1987
1988-89	James Watson, Director, NCHGR of NIH	Political support became available for sequencing complete human genome
1990-91	Daniel Cohen (CEPH) and Jean Weissenbach (GENETHON)	Map first and sequence-later strategy was used and high resolution map constructed
1992-95	Francis Collins, Head, NCHGR (Watson had problems with Craig Venter of NIH)	Craig Venter sets up The Institute for Genome Research (TIGR) in Reckville (Maryland), which sequenced several bacterial genomes
1996	A meeting of international partners in Bermuda	Conditions of data access (including release of sequence into public databases) were formalized — "Bermuda Principles"
1998	Craig Venter formed a company 'TIGR', later named 'Celera Genomics'	Decision to sequence the human genome in three years using 'whole genome shotgun' (WGS) approach without using maps.
1999	Five centres in USA/UK start bulk sequencing	Public project decides to produce a draft genome sequence by 2000, in view of Craig Venter's challenger; chromosome 22 sequenced
2000-2001	HGP and Celera Genomics joined and completed the first drafts of human genome	Chromosome 20 and 21 were sequenced and draft genome sequence was released; public sequence was published in Nature and Celera's sequence published in Science
2001-2004	A number of centres in USA and UK	Chromosomes 6, 7, 9, 10, 13, 14, 19, 22 and Y were fully sequenced by May 26, 2004. These finished sequences are 99.9% accurate

The draft prepared by the International Human Genome Sequencing Consortium (IHGSC) was published in Nature (in UK) and that prepared by Celera Genomics was simultaneously published in Science (in USA). Thousands of scientists around the world worked round the clock for 15 years to achieve this feat, which was compared by some with man's landing on the moon in 1969. Rightly, according to a scientist involved in this work, it gave us the feeling of being on a satellite and looking down at earth with excitement. Later, by the end of May, 2004, as many as 11 human chromosomes were fully sequenced with an accuracy of 99.9%. The major events that followed human genome sequencing efforts, starting in the year 1985, upto 2004 are chronologically listed in Table.

Some Details of Chromosomes 20, 21 and 22, that were Fully Sequenced by the end of 2001.

Features	*Human chromosome that was sequenced*		
	20	21	22
Institute	Sanger Institute	Consortium	Consortium
Length sequenced	63.3 Mb	33.82 Mb	33.79 Mb
Genes	727	225	545
Pseudogenes	168	100	134
CpG islands	660	115	553
SNPs	29,478	20,916	28,410
Diseases associated	Crenzfeldt Jakob, ADA	Down syndrome	Cat eye syndrome (CES)

Human Chromosomes Fully Sequenced : Although the rough drafts of the entire genome sequence were published in February, 2001, as many as eleven chromosomes (chromosomes 6, 7, 9, 10, 13, 14, 19, 20, 21 and 22, Y) were fully sequenced by May, 2004. Some details about three chromosomes are presented in Table.

Bacterial Genes Transferred to Human Genome : In the rough draft sequence of human genome, it has been found that during evolution, roughly 113 to 223 genes have been laterally transferred from bacterial genome to human genome, a phenomenon that is described as horizontal gene transfer (HGT); these are also described as bacteria-to-vertebrate transfers (BVTs) and their number according to some other estimates is much smaller than the above estimates. The hypothesis of lateral transfer of genes was based on similarity searches, where vertebrate genome sequences were searched for bacterial genes that are absent in other eukaryotes. The endosymbiont bacteria that gave rise to chloroplasts and mitochondria could be one source of these lateral transfers.

Mouse (Mus musculus) Genome : In mouse, the haploid chromosome number, $n = 20$ is smaller than the corresponding chromosome number, $n = 23$ in humans. Despite this, the genome size in mouse is either equal to or slightly bigger (3,300 Mb) than that in human (3,200 Mb), although the euchromatic part of the mouse genome is 14% smaller (2,500 mb) than the corresponding euchromatic part of the human genome (2,900 Mb). Efforts have been made by both public and private sectors to sequence the mouse genome. As a result, a rough draft sequence of the mouse genome was released by private sector (Celera Genomics) in April 2001, and a high quality draft sequence was published in December, 2002 (Nature, 5 December, 2002). The final precise and accurate sequence of mouse genome should become available by the year 2005.

During the year 2001, using the human genome as a framework map, a physical map of the mouse genome was also completed by four groups, which included Sanger Centre, near Cambridge (England), the Genome Sequence Centre in British Columbia (Canada), Washington University in St Louis, Missouri (USA) and the Institute for Genome Research (TIGR) in Rockville, Maryland (USA). This physical map, available at the Ensembl database, later provided a complete tiling path for each chromosome and facilitated anchoring of the genomic sequences generated by the International Mouse Genome Sequencing Consortium (MGSC). In this map, 290,000 BAC clones were assembled into just 554 contiguous blocks (contigs) spanning 3,080 megabases, representing 98% of the entire genome. This was a small number of contigs, which made it easy to assemble the sequence of the whole genome by using the shotgun sequencing (WS) approach that was pioneered by Celera Genomics of Rockville. This involved sequencing of 6 million fragments, ranging in length from 500 to 700 base pairs.

The draft mouse genome sequence that was released in 2001 belonged to Celera Genomics and was not freely available. However, on December 5, 2002, Mouse Genome Sequencing Consortium (MGSC) reported a high quality mouse genome draft sequence which was the result of international collaboration. This draft sequence for mouse genome has been used and will continue to be used in future for comparison with human genome sequences and will thus help in unlocking some of the secrets of human genome. Some of the similarities and differences between the human and mouse genomes include the following :

(i) Ninety Per cent of the two genomes exhibit conserved syntony in segments of genomes although at the nucleotide level only 40% of the human genome can be aligned to the mouse genome;

(ii) Each of the two genomes seems to contain about 30,000 protein coding genes; 80% of mouse genes have a single orthologue in human genome and the proportion of genes having no orthologue in other genomes is less than 1%,

(iii) After the evolutionary divergence of the two genomes on geological time scale, nucleotide substitutions in mouse genome are twice as many as in the human genome.

A 'Mouse Gene Encyclopaedia Project' was also initiated in the year 2001 to determine the full coding potential of the mouse genome. An international functional annotation meeting (FANTOM) was also organized to annotate the cDNAs that were analysed by the project. In Phase I of the project, 160 full length enriched and subtracted cDNA libraries were prepared from various tissues and developmental stages. From these cDNA libraries, 930,000 3' end sequences were collected and clustered into 128,000 groups. In phase II of the project single clones from individual clusters were selected for sequencing, although all clusters were not represented in 21,076 cDNA clones (average length, 1,257 bp) that were sequenced. These sequenced clones were tested for similarity to mouse genes represented in the database of Mouse Genome Informatics (MGI), and only 4,248 clones were found to be identical to mouse genes. After annotation, the clones were classified into different categories of genes (15,296 genes); more than five thousands of these genes were still in the unclassified categories.

Rat (Rattus Norvegicus) Genome : Rat genome, measuring almost as much as the human/mouse genome is the third mammalian genome that has been fully sequenced in the year 2004. The project received support of US $ 60 million sanctioned by US Government to three laboratories (Baylor College of Medicine, Houston, Texas; Celera Genomics, Rockville, Maryland and Genome Therapeutics Corp., Waltham, Massachusetts). Brown Norway (BN) rat has been utilized for the rat genome sequencing project. Rat genome sequence was considered important, because rat has been used more often than mouse for the study of physiology in mammals. It is believed that having sequences of the genomes of three most important mammals (human, mouse and rat) for biomedical research will help in unravelling the genetics and physiology of all human diseases with great speed.

Chimpanzee and Macaque Genomes : Once the draft human genome sequence became available in February, 2001, primate experts started lobbying to get the genomes of chimpanzee *(Pan troglodytes)* and rhesus monkey or macaque *(Macaca mulatta* sequenced. A draft sequence of chimpanzee genome was made available on the internet in April/May, 2004 and a comparison of chimpanzee and human genomes should be published in late 2004. It is argued that while chimpanzee genome sequence should give us a clue about what makes us human, macaque genome sequence should facilitate the study of different human diseases, and the testing of new vaccines and drugs. Taxonomists, however, argue that from phylogenetic perspective, these are not the best choice.

Pufferfish and Zebrafish Genomes : Pufferfish *(Takifugu nigroviridis* or *T. rubripes)* is a vertebrate with a small genome (400 Mb), which is less than one seventh the size of the human genome, but contains approximately the same number of genes. The pufferfish genome, therefore, was selected for whole genome sequencing. The mapping carried out with pufferfish genome indicated that there is some similarity with the human gene order. This means that sequencing of pufferfish may help in finding human homology of pufferfish genes. Such an exercise may prove useful in locating undiscovered genes in human genome. The greatest promise which pufferfish genome sequencing offers is the possible identification of essential promoter and other regulatory sequences that lie upstream of human gene sequences and may be similar in pufferfish and human genomes.

Zebrafish *(Danio rerio)* is another vertebrate with a genome size of 1.7×10^9 bp, which is about half the size of human or mouse genome, which is 3.2×10^9 bp. A zebrafish is 4 cm long, its female laying 200 eggs per week and its transparent embryos reaching maturity within 2-3 months. Being a vertebrate, zebrafish has a heart, blood, hearing system, kidney, cartilage, bone and optical systems that share many features of the human systems. This is unlike the nematode, *C. elegans* and the fruitfly, *Drosophila melanogaster*, whose genomes are fully sequenced, but which do not possess many of the complex organ systems found in vertebrates. In view of this, it was believed that sequencing of the genome of zebrafish will complement the information on genome sequencing available for human, mouse and rat. Another great advantage of using zebrafish as a model vertebrate for genome sequencing is its ability to generate readily a large number of mutations that are relevant to human health and disease. More than 1000 zebrafish mutations that were already available in the year 2001, are known to affect processes, which have great relevance to human physiology and diseases. The number of these mutations will increase in the early years of the 21st century.

In February, 1997, the zebrafish community had organized a workshop entitled 'Current Advances in Defining the Zebrafish Genome'. This workshop made a recommendation that zebrafish be promoted as a model organism for the study of vertebrate development and diseases. Consequently, towards the end of 2001, NIH (USA) established the trans-NIH zebrafish coordination committee, to formulate an NIH Zebrafish Initiative, which also has a web site. As a result of this new initiative, molecular map of zebrafish genome with ~ 400 molecular markers (25 linkage groups) became available in 2002. Wellcome Trust and Sanger Centre (UK) also recommended that after human/mouse/rat genome, the next target for genome sequencing should be zebrafish, which was started in February 2001 and was estimated to take three years. Sequencing will be done using both clone by clone approach and whole genome shotgun (WGS) sequencing. Sequencing of this genome will provide a rapid route to the discovery of molecular basis of available mutations. The genome of zebrafish will be fully sequenced by the end of 2004 and will be made freely available to researchers without any cost or restriction.

Genomes of Other Animals : In addition to the genomes of human, mouse, rat, pufferfish and zebrafish discussed above, genomes of several other animals are planned to be sequenced in the early years of the 21st century. For instance, in the year 2001, representatives from 20 research centres in 12 countries initiated a plan for sequencing the genome of *Anopheles gambiae* (260 million bp = 1/12th of human genome), the main vector of the malarian parasite in sub-Saharan Africa. The genome of the malarian parasite *(Plasmodium falciparum)*, which kills 1.5 million children every year, was already sequenced in 2002 (Nature, October 3, 2002). The main sequencing of *Anopheles* genome was undertaken at Genoscope (French sequencing centre) in France and Celera Genomics in USA, using the whole-genome shotgun (WGS) sequencing approach that was used by Celera Genomics for human genome sequencing. The *Anopheles* genome was also fully sequenced in 2002 and the results were published on October 4, 2002 in Science.

genomes of several other animals are planned to be sequenced in the early years of the 21st century. For instance, in the year 2001, representatives from 20 research centres in 12 countries outlined a plan for sequencing the genome of *Anopheles gambiae* (260 million bp = 1/12th of human genome), the main vector of the malarial parasite in sub-Saharan Africa. The genome of the malarial parasite (*Plasmodium falciparum*), which kills 1.5 million children every year, was already sequenced in 2002 (Nature, October 3, 2002). The main sequencing of *Anopheles* genome was undertaken at [illegible]. The main sequencing [illegible] centre [illegible] Genomics, USA, using the whole genome shotgun (WGS) sequencing approach that was used by Celera Genomics for human genome sequencing. The *Anopheles* genome was also fully sequenced in 2002 and the results were published on October 4, 2002 in Science.

Plant Genomics

The detection and exploitation of naturally occurring DNA sequence polymorphisms represent one of the most significant developments in molecular biology. Earlier, we briefly described the techniques of development of molecular markers and those for the whole genome sequencing. The molecular maps and complete sequences for some animal genomes were also described in that chapter. The preparation of similar molecular maps and determination of whole genome sequences in crop plants has been a priority area of research in plant biotechnology also. The utility of molecular maps for crop improvement programmes has been emphasized during 1980s, so that the molecular maps have already been prepared for all major crop species. More recently YAC and BAC physical maps have also been prepared and used for 'clone to clone' whole genome sequencing. Whole genome shotgun has also been utilized in rice genome sequencing and the technique will certainly be used for sequencing of the genome of other crop plants. These aspects of plant biotechnology, including construction of molecular maps and whole genome sequencing, will be briefly discussed in this chapter.

Preparation of Genetic Maps

In the preparation of molecular genetic maps, following steps are involved : (i) development of a mapping population, (ii) selection of one or more molecular marker systems, (iii) identification of adequate number of molecular markers showing polymorphisms in the mapping population, and genotyping of individual plants or lines of the mapping population, using each polymorphic molecular marker, and (iv) construction of molecular map using suitable computer software.

Mapping Populations and Near Isogenic Lines (NILs) : A mapping population for the preparation of a genetic map should consist of a number of plants or lines, which should be the products of one or more meioses. Therefore, such a population need to be derived from a cross between two inbreds or pure lines, which should be diverse.

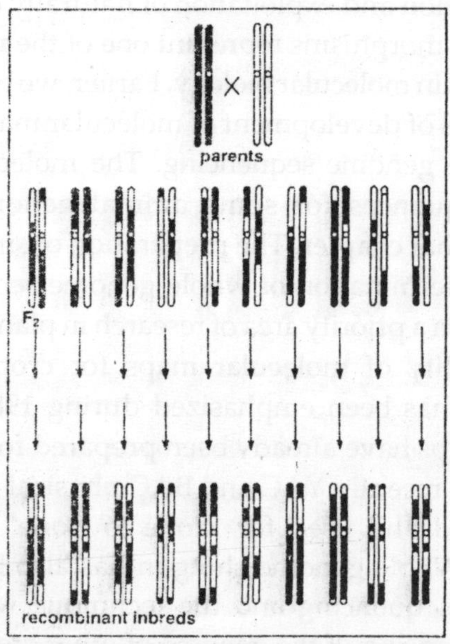

The production of recombinant inbred lines. This figure depicts a hypothetical diploid organism with a single pair of chromosomes. The original parents are themselves inbred lines and homozygous at all loci. Individuals of a large F_2 population produced from a mating of these inbreds are self-pollinated, in case of maize for at least six generations, to produce new recombinant inbred lines (RILs).

The mapping population may, therefore, consist of either an F_2 population, or a backcross (BC_1) population, or a doubled haploid (DH) population each derived from F_1 plants or a set of recombinant inbred lines (RILs) generated using single seed descent (SSD) method of breeding (Fig.). Such a population can be mortal, as in case of F_2 or BC_1, or immortal as in case of DHs or RILs.

List of Available Recombinant Inbred Population in maize.

Parents	*No. of RI lines*	*Generation of inbreeding*[a]
T232 × CM 337[b]	48	8
CO159 × T × 303[b]	41	8
CO 159 × T × 303	160	5
B73 × Mol (IBM*)[b]	44	7
PA326 × ND300	74	5
CK52 × A671	162	5
CG16 × A671	172	5
CH593 – 9 × CH606 - II	101	5
CO220 × N28	173	5

Generations of selling beyond the F_2; *IBM = intermated BM [b]Extensive mapping has been done using these families.

Although mortal F_2 and BC_1 populations have successfully been utilized for the preparation of molecular genetic maps in a variety of crop plants, an immortal DH or RIL population is preferred, since it can be repeatedly utilized not only for mapping, but also for a variety of other purposes including gene tagging and QTL analysis for important economic traits (QTL = quantitative trait loci). A few sets of maize RILs that have been used as mapping populations are listed in Table.

The backcross programme is also utilized for generation of near isogenic lines (NILs) for specific traits, where a pair of lines (NILs) differ for only small segment(s) of one or more chromosome(s) carrying gene(s) for the concerned trait. In doing so, an initial cross is made between a recurrent parent (often a desirable genotype deficient for a particular trait) and a donor parent (carrying the gene of interest, although otherwise agronomically poor). The F_1 plants are backcrossed repeatedly with the recurrent parent to reconstitute its genotype. While making backcrosses, in each generation only selected plants carrying the gene of interest are used.

Different Molecular Marker Systems : Different molecular marker systems including RFLP, RAPD, SSR and AFLP, have been used for the preparation of molecular maps in several crops. However, the latest of these markers called single nucleotide polymorphisms (SNPs), which have already been used for the preparation of a molecular map of human genome, could not be used so far (till 2004) for the construction of a map of complete genome in any crop plant, although an SNP map for *Arabidopsis* genome was published in 1999. But SNPs have already been used for the detection of polymorphism in crop plants suggesting that this new class of molecular markers will also be used in future for preparing molecular maps in crop plants.

Polymorphic Markers and Genotyping the Mapping Population : Before molecular markers are mapped, they need to be tested for detecting polymorphism between the two parents of the mapping population. For further mapping work, only those markers are used, which exhibit polymorphism between the parents of mapping population, since markers showing no polymorphism can not be mapped genetically (although they can be mapped physically; see later for physical maps). The two polymorphic states of the marker recorded in two parents are given allelic symbols or 1 and 0 (zero) states, and all individuals/ lines (~100) of the mapping population are genotyped (assigned specific allele or 1/0 state). Sometimes, these allelic states are called A and B. For codominant markers, like RFLPs, if mapping population belongs to F_2 generation, there will be three genotypes (AA, aa, Aa), which can be scored as 0, 1 and 2 (or A, B and H) to be used with the computer software. The data on genotyping of the mapping population, involving molecular markers ranging in number from a few hundred to few thousand, is then used for the preparation of a genetic map, which is based on recombination frequencies that are automatically worked out by the software and utilized in a mapping function for determining genetic distances.

Preparation of Map using the Software MAPMAKER : Keeping in view the amount of data (~100 individuals/lines, each genotyped with few thousand markers) to be handled for the preparation of a molecular genetic map, computer softwares are used. One such software in common use is MAPMAKER, which makes use of either Haldane's mapping function (no interference is assumed) or Kosambi's mapping function (interference is taken into consideration). Other softwares like LINKAGE were used earlier for the same purpose.

Molecular Genetic Maps (Low/Moderate Density) : During the last two decades (1980-2000) and in the early years of the 21st century, molecular genetic maps became available for all major crop and tree species (cereals, legumes, oil seeds, vegetables, trees, etc.). The mapping populations that were used for the construction of these maps included F_2, BC_1, RIL and DH populations, and the molecular markers that have been used for this purpose mainly included RFLP, RAPD, SSR and AFLP, although other marker systems like STS, ISSR and DAF were also utilized (consult author's book *"Biotechnology and Genomics"* for more details on molecular markers). For constructing a mapping population, when sufficient variability was not available within the germplasm of the crop, a wild species had to be crossed with a cultivar to provide for the diversity, which is a key to the success of constructing a map. For instance, in case of tomato, a wild species, *Lycopersicon hirsutum* or *L. pennellii* was often used; in case of barley, *Hordeum spontaneum* (wild barley) was used, and in case of bread wheat, a synthetic wheat (W7984) involving *Aegilops tauschii*, the D genome progenitor, was used for crosses leading to the construction of one or more mapping populations. In sugarcane, *Saccharum spontaneum*, the wild relative of sugarcane was used either directly for the preparation of a molecular map or for crosses with cultivated species *S. officinarum* for generating a mapping population.

In each of the majority of crops used for molecular mapping, more than one molecular maps are often available. This is because, either separate maps were prepared using different marker systems, or more than one laboratories prepared independent maps, utilizing the same marker system, thus multiplying the number of maps for the same crop. This multiplicity of maps was sometimes necessary, because often in a single population, only a fraction of available markers segregated, so that several populations were needed to map all the available markers. In some crops, eventually useful integrated maps were also prepared, utilizing the information from all the available maps (see later). Some details about the available maps are presented in Table where it can be seen that most of these maps are low to moderate density maps with variable resolution, suggesting that in many cases these maps still need to be saturated (see next section for high density maps). The present status of these molecular maps (at the beginning of the present century) in different crops will be briefly discussed in this section. For any details, readers should use the internet, where more detailed information about most of these maps is readily available.

Maps for Cereals, Millets and Sugarcane (grass family) : Molecular maps have been prepared in all major cereals, millets and other graminaceous crops including maize, bread wheat, emmer wheat, oats, barley, rice, sorghum, pearl millet, sugarcane, etc. The most extensive maps among cereal genomes, however, are available in case of rice, which has become a model for all graminaceous crops and has therefore been the first food crop to be used for genomics research (due to its small genome size of 430 Mb).

Details of Molecular Maps in Some Crops

	Gametic chrom. no.	No. of marker loci	Map length (cM)	Genome size (Mbp)	Resolution bp/cM
I. Poaceae					
Maize *(Zea mays)*	10	-4800	>1000	2500	2.5×10^6
Rice *(Oryza saliva)*	12	-3300	>1800	430	0.29 × 10*
Bread wheat *(Triticum aestivum)*	21	4700	>5000	16000	3.2×10^6
Emmer wheat (T. *dicoccoides)*	14	3000	3000	-12000	4.0×10^6
Barley *(Hordeum vulgare)*	7	2000	1500	5200	3.5 * 10"
Sorghum *(Sorghum bicolor)*	10	-2500	-1050	750	0.71×10^6
II. Solanaceae					
Tomato *(Lycopersicon esculentum)*	12	3000	1500	950	0.67×10^6
Potato *(Solarium tuberosum)*	24	>6500	1100	-2000	1.8×10^6
Pepper *(Capsicum annuum)*	12	>2200	1800	-3900	2.7×10^6
Eggplant *(Solanum melongena)*	12	-250	-1500	-950	0.63×10^6

Note : - = ~

Contd....

	Gametic chrom. no.	No. of marker loci	Map length (cM)	Genome size (Mbp)	Resolution bp/cM
III. Leguminosae					
Chickpea *(deer arietinum)*	8	500	-2000	1000	0.5×10^6
Pea *(Pisum sativum)*	7	850	-1500	4200	2.8×10^6
Common bean *(Phaseolus vulgaris)*	11	1000	-1000	600	0.6×10^6
Soyabean *(Glycine max)*	20	>1800	-2500	1100	0.4×10^6
Alfalfa *(Medicago sativa)* - diploid	8	-870	-750	-800	1×10^6
Alfalfa *(M. saliva)* - tetraploid	16	-90	-450	1700	$3.8 * 10^6$
Lentil *(Lens culinaris)*	7	200	500	4100	8.2×10^6
Lotus japonicus	6	-600	-400	-450	1.12×10^6
M.truncatula	8	550	1000	-500	0.5×10^6
IV. Cruciferae					
Brassica oleracea	9	550	900	700	0.78×10^6
B. napus	19	-1000	2500	-1300	0.52×10^6
B.juncea	18	100	1600	-1500	0.93×10^6
V. Malvaceae					
Cotton *(Gossypium hirsutum)*	26	-3300	>5500	-2200	0.4×10^6
VI. Compositae					
Sunflower *(Helianthus annuus)*	17	-270	-1600	-2400	1.5×10^6

As many as about 2300 marker loci spanning a genetic distance of 1800 cM, have been mapped in this crop. Sorghum is another important crop that has been targeted for plant genomics research, because it is adapted to harsh environments, and is represented by diverse germplasm and a small genome (750 Mbp). In this crop about ten molecular genetic maps with moderate density of RFLP and SSR markers spreading over a genetic distance of up to -1800 cM were prepared during 1990s (for high density maps in rice and sorghum, see next section). In bread wheat, RFLP maps were initially prepared by joint efforts of different laboratories, which were co-ordinated by the International Triticeae Mapping Initiative (ITMI). Subsequently, microsatellite (SSR) maps in bread wheat were also prepared by several laboratories in different parts of the world. Molecular maps with low to moderate density were also prepared in barley, rye, oats, pearl millet, sugarcane, etc. For instance, barley genetic maps were prepared independently in Europe and USA. An SSR map of barley was also prepared in the year 2000, using 242 of more than 600 SSR primers that were available in the public domain at that time. Molecular maps based on RFLPs and AFLPs have also been prepared for diploid and hexaploid oats. In sugarcane, both cultivated and wild species of *Saccharum* were used for preparation of molecular maps involving RFLPs and RAPDs. Detailed information on majority of cereal molecular maps is available on GrainGene and Gramene sites, although web sites are also available for individual crops (e.g. for maize, maizeDB:<www. agron.missouri.edu> and for sorghum, <http: / / sorghumgenome.tamu.edu>).

Maps in Legumes (Soyabean, Pea, Common Bean, Alfalfa, Chickpea, Lentil) : All major legumes have been used for constructing molecular maps, although a moderately dense map is available only in case of soyabean, where more than one thousand loci involving all kinds of markers (RFLP, RAPD, SSR, AFLP) were mapped over a length of ~ 2500 cM. In pea also, a fairly good map involving ~ 850 marker loci spanning a genetic length of ~1500 cM became available towards the end of last century. Similarly, in common bean *(Phaseolus vulgaris)*, a map with as many as 1000 RFLP and RAPD markers is available. In contrast to these moderate density maps in soyabean, pea and common bean, the available maps in chickpea and lentil are sparse and low density maps, with about 500 mapped markers in chickpea and 200 mapped markers in lentil Table. Maps have also been prepared in some other legumes including *Lotus japonicus (2n* = 12), alfalfa *(Medicago saliva)* (*2n* = 32) and the model legume, *M. truncatula* (*2n* = 16).

Maps in Brassicas : Molecular maps have been prepared in a number of *Brassica* species including both (i) the diploid Brassicas like *B. oleraceae* (2n = 18) and *B. nigra* (2n = 16), and (ii) the polypoid Brassicas like *B. napus (2n* = 38) and *B. juncea (2n* = 36). In these crops, mainly RFLP, RAPD and AFLP markers were used, since SSR markers were not available in large numbers for mapping. For instance, in *B. oleracea*, 258 RFLP markers were mapped in 1990 and later an integrated map with 543 marker loci was published in 2001. A skeleton map with 67 RAPD loci was also prepared in *B. nigra*. Similarly, in *B napus*, a consensus map with 542 RFLP loci and 253 AFLP loci was published in 2001 and a dense map in *B. juncea* (Indian mustard) with 1029 loci (996 AFLP and 33 RFLP) was published in 2002 from a group from Delhi University South Campus (UDSC), India.

Maps in Cotton : A number of molecular genetic maps have been prepared in cotton, most of them for tetraploid upland cotton *(Gossypium hirsutum)* with AADD genomic constitution, but some of them also for diploid cotton (*G. herbaceum)* with AA genomic constitution. The mapping populations used for mapping the genome of tetraploid cotton, generally consisted of F_2 or $F_{2/3}$ (bulk-sampled plots) or DH lines derived from inter-specific cross *G. hirsutum* × *G. barbadense*. Some of these maps included the following : (i) In

1998, a detailed restriction fragment length polymorphism (RFLP) map, was prepared for tetraploid cotton, in which 705 RFLP loci, assembled into 41 linkage groups, covered a genetic distance of 4675 cM. The genomic origin (A *vs* D) of most, and chromosomal identity of 14 (of 26) linkage groups was shown. (ii) Another molecular linkage map was constructed using 58 doubled haploid plants derived from the inter-specific cross *G. hirsutum* × G. *barbadense.* Among a total of 624 marker loci (510 SSRs and 114 RAPDs), 489 loci were assembled into 43 linkage groups and covered 3,314.5 cM. (ii) A joinmap from four populations comprising 284 loci spanning a genetic distance of approximately 1,500 cM was also published in 2001. This was the first genetic linkage joinmap assembled in G. *hirsutum* with a core of RFLP markers assayed on different genetic backgrounds of cotton populations (Acala, Delta, and Texas plain). These maps taken together also demonstrated that there are duplicate loci for many markers due to homologous relationships between the two genomes (A and D) that are present in tetraploid cotton.

***Maps for Sunflower* :** Several molecular maps have been prepared in sunflower. The molecular markers that have been used for this purpose included RFLPs, SSRs, DALPs and AFLPs. These maps, each consisted of 17 or 18 linkage groups ($n = 17$) and differed in their genetic length from 1368 cM in the SSR map to 2168 cM in DALP/AFLP map. The latest of these maps included an AFLP map published in 2001 and an SSR map published in 2002. Numerous duplicated loci were detected by many RFLP probes, suggesting that sunflower is a secondarily derived polyploid.

***Maps for Solanaceous Crops* :** Solanaceous crops that were used for molecular mapping, mainly included tomato ($2n = 24$), potato (2n = 48) and pepper ($2n = 52, 104$). Information on these maps is freely available on SolGenes database. A molecular map of tomato was perhaps the first molecular map that was prepared in plants and should now contain more than 3000 mapped loci.

Maps for Forest Trees and Fruit Trees : Molecular maps have also been prepared in several tree species that are important for forestry. These tree species include species from both gymnosperms and angiosperms. The gymnosperm trees that have been used for genome mapping mainly include a number of pines {e.g., loblolly pine *(Pinus taeda)*}, radiata pine *(Pinus radiata)*, slash pine *(Pinus eliottii)*, longleaf pines *(Pinus palustris)*, martime pine *(Pinus pinaster)*, scots pine *(Pinus sylvestr:s)*}, at lcast two spruces {white sprue *(Picea glauca)* and Norway spruce *(Picea abies)*} and Douglas fir *(Pseudotsuga menziesii)*. Similarly, angiosperm trees, whose genomes have been mapped include European larch *(Larix deciduas)*, white birch *(Betula alba)*, poplar *(Populus sp)*, and several species of *Eucalyptus* and *Quercus* (oaks). A variety of DNA-based molecular markers that have been used for preparing these maps include RFLPs, RAPDs, AP-PCR, DAF, STS, SSLPs, ESTPs, SCARs, etc.

Among fruit trees, genomes of apple *(Pyrus malus*, syn. *Mains sylvestris)*, apricot *(Prunus armeniaca)*, peach *(Primus persica)*, lemon *(Citrus* sp.), avocado *(Persea americana)*, etc. have been subjected to molecular mapping. In some cases (particularly in case of apple), permanent mapping populations have been established for mapping. As in other cases, the molecular markers that have been used for mapping of fruit trees include RFLPs, RAPDs, SSRs and AFLPs. In both, forest trees and fruit trees, the mapped molecular markers are being used for gene tagging and QTL analysis, so that in forest and fruit trees also, molecular markers will be utilized in future for marker assisted selection and gene isolation.

Molecular Genetics Maps (High Density) : Although molecular genetic maps are now available in majority of crops, most of these maps are inadequate for identification of candidate genes and for subsequent map based cloning of genes. Therefore, efforts have been under way to prepare high-resolution maps in several crops. One of the approaches followed in maize in this connection involved intermating among plants in F2 and subsequent generations to provide for an opportunity for increased recombination. When such a population, after five-generations of intermating, was used in maize, it led to the production of a high resolution map, in which the map distance increased from 1532 cM to 5917 cM. Intermated RILs were also used in maize to prepare a high density SSR map involving 978 SSRs. In sorghum also, high density map with ~3000 marker loci spread over 3000 cM was later produced during 2000-2002. A high-density genetic linkage map of tetraploid cotton was also developed, which had 2662 loci (RFLPs involving 75% cDNA and 25% gDNA, RFLPs; 5% SSRs), which were distributed in 26 linkage groups and spanned a genetic distance of over 4700 cM. An integrated high density map in *Aegilops tauschii* genome, which represents the D genome of bread wheat, was also published in 2002; this map had 732 loci. In several crops including bread wheat and maize, efforts have also been made to identify gene-rich genomic regions (these regions are also shown to be the recombination hot-spots) in the maps. These gene rich regions are being subjected not only to high resolution mapping but also to genome-wide sequencing.

Integrated Genetic Maps

Integrated maps have also been prepared utilizing multiple maps involving either the same class of markers, or more than one marker classes. For instance, integrated maps have been prepared by integrating microsatellite maps into RFLP maps in several crops including rice, wheat, barley, maize, soyabean, potato, tomato, sugarbeet, etc.

Resolution Gap Differs in Maps with Genomes of Different Sizes : In preparing genetic RFLP maps, lcM (centiMorgan) is often considered to be the limit of resolution achievable without examining a very large population. At the molecular level, this limit of resolution means that 1 cM would be equal to different physical distances in organisms, with genomes of different sizes, as shown in Table. This gap, therefore, needs to be bridged, if a gene of interest is to be approached from a linked RFLP locus, that has been mapped. Bridging of this gap has become possible due to development of techniques like pulse field gel electrophoresis (PFGE) and cloning in yeast artificial chromosome (YAC) vectors, which will enable handling of large pieces of DNA. Similarly jumping and linking libraries can be prepared, so that chromosome walking need not be followed throughout the genome to identify, locate and isolate a gene of interest.

The Resolution Gap as Worked out in Four Different Genomes.

Organism	*Genome size (kbp)*	*Map length (cM)*	*kbp equal to 1 fM*
Arabidopsis	12.5×10^4	~500	250
Tomato	9.5×10^5	1500	670
Maize	2.5×10^6	1000	2500
Human	3×10^6	2700	1111

Uses of Molecular Genetic Maps

Mapping of Major Genes of Economic Value using Molecular Markers (tagging) : Molecular markers can be used to map positions of genes for simple traits as well as those for polygenic traits (also described as quantitative trait loci or QTLs). Once these genes are mapped, the molecular markers linked to specific genes or gene clusters (gene tagging) can be used in selection schemes for plant breeding (see later for details). Molecular markers mapped very close to important genes can also be used for isolation of these genes through map-based cloning.

Use of Near Isogenic Lines (NILs) for Gene Tagging : In recent years, near isogenic lines (NILs) have been utilized to identify linkages between molecular markers and conventional phenotypic markers. An NIL is produced, when a gene for conventional phenotypic marker is transferred from a donor parent (DP), into a recurrent (RP) and the genotype of the recurrent parent is restored through 5-7 backcrosses, retaining the conventional phenotypic marker in each such backcross. Therefore, allelic contrast between RP and its NIL, and allelic equality in the corresponding DP and NIL for a molecular marker suggests linkage of molecular marker with the conventional phenotypic marker.

This, however, needs to be confirmed through F_2 cosegregation data generated from NIL × RP crosses. This will also give an estimate of the linkage distance between the conventional phenotypic marker and each of the associated molecular markers. This technique has been successfully employed for tagging genes of economic value in several crops including, tomato, potato, maize, barley, rice and sugarbeet. Hundreds of cases of such linkages of molecular markers with genes for economic traits in a variety of crop plants are now known.

Identification and Mapping of Quantitative Trait Loci (QTLs) using Molecular Markers : Many quantitative traits of economic value are under polygenic control and are selected for, directly. Such a selection is often ineffective, since the effect of each gene is small, which is also influenced by the environment. Therefore, one would welcome a procedure for indirect selection, which is not influenced by the environment. For this purpose, linkage between genes for quantitative traits and marker loci should be known. Such markers involving morphological traits have been identified and mapped in several crop plants. Isozyme marker loci linked to quantitative traits of economic value had also been earlier identified in tomato and maize. Initially, RFLPs were extensively used in a variety of crop plants to identify and map quantitative trait loci (QTLs). More recently (1995-2004), SSR and AFLPs became the markers of choice and were widely used. In future, SNPs will also be used in many crops.

(a) *Comparison of Means of Marker Classes* : For the study of marker-trait association, the experimental procedures differ for self pollinated and cross pollinated species, although with the availability of inbred lines in cross pollinated species, similar methods can be used in the two cases. Among pure lines (in self pollinated species) and inbred lines (in cross pollinated species), parents are first selected which differ for a number of markers as well as for mean values of the quantitative traits. Once such parental lines are selected, they are crossed and F_2 plants are derived. In the F_2 population, if mean values of a particular quantitative trait are determined in two groups of plants representing alternative alleles of a molecular marker, a significant difference in means (of two groups) for quantitative trait will indicate marker-QTL linkage. The observed marker associated difference in quantitative trait value will be a function of allelic effects at the QTL. Map distance between the marker and a QTL can also be determined.

In the Table results of an initial experiment are shown where in tomato, linkage is evident between one of the two molecular marker *(A_1, A_2)* and a QTL for high soluble solids in the fruits. Two lines (cultivars VF 36 and its near isogenic line LA 1563) used in the cross had molecular markers (RFLPs) that identified *L. esculentum* (~5% soluble solids) and *L. chmielewskii* (~10% soluble solids), and also differed significantly for soluble solids (~5% and 7-8%). In F_2, allele 'C' of the marker A_1 showed association with the QTL for high mean value of soluble solids, while the allele *'e'* of A_1 and both the alleles of A_2 did not, suggesting that this QTL is found in the vicinity of A_1. RFLPs, associated with the expression of complex traits like insect resistance and WUE, i.e. water use efficiency, were also identified and mapped in tomato. Subsequently, thousands of such marker-trait associations were reported in a number of crops. For instance, more than one hundred such associations have been reported each in wheat and rice. In the experiments like the above, if parents differ for a sufficient number of molecular markers, which are spaced out throughout the genome, a single cross can allow identification and mapping of all those QTLs for which the parents differ.

The Data on Soluble Solids Content (mean per cent Values) and two RFLPs in a F_2

(F2 was derived from a cross between a tomato cultivar (~5% soluble solids) and its near isogenic line (NIL) carrying a QTL for high (7-8%) soluble solids derived from L. chmielewskii with 10% soluble solids; data from Osborn et. at. 1987).

RFLP locus	*% Soluble solids*	
	RFLP allele e/e (*L. esculentum*)	RFLP allele c/c (*L. chmielewskii*)
A_1	5.82	6.24*
A_2	5.93	5.96

*significant difference at 5% level.

(*b*) *QTL Interval Mapping* : New analytical methods for the analysis of quantitative trait loci in plants using mapped molecular markers were also suggested. These new methods included maximum likelihood or regression method for interval mapping which was earlier used in humans. Interval mapping assesses the effects of small genome segments (each~2cM), located between a pair of marker loci, rather than the effect of a QTL associated with a single RFLP. This provides greater precision. LOD scores (log of the score of odds for the presence of a linkage against the odds of not having this linkage, e.g. if the score is 100 : 1 in favour of linkage, the LOD score is 2) are also used in this method to reduce the probability of false positives for linkage (detecting linkage, when it is actually absent). The results are depicted as 'QTL likelihood maps'. The technique of interval mapping allowed the QTLs to be mapped to intervals of 20 cM which was a poor resolution. Therefore, the technique of interval mapping was supplemented through the use of substitution mapping using selected overlapping recombinant chromosomes. This technique is analogous to the use of overlapping deletions and allowed fine mapping of QTLs. The method has been successfully used for a study of tomato fruit characters, including mass per fruit, soluble solid concentration and fruit pH. The QTLs for these characters were mapped at intervals, as small as 3 cM.

Simple interval mapping (SIM) that was proposed initially by Lander and Botstein in 1989, and used subsequently in early 1990s suffered with some serious problems. For instance, it allowed identification of one QTL at a time, ignoring the effect of other QTLs in the background of the concerned genome. Furthermore, it was not possible to determine the precise location and number of QTLs; nor was it possible to estimate the non-allelic QTL × QTL interactions (epistasis) and the QTL × environment interactions. In view of these limitations of SIM, modified forms of interval mapping like composite interval mapping (CIM) and multiple interval mapping (MIM) were suggested and are now being increasingly used for QTL analysis. These modified forms take into account the QTLs that are present in the background and allow estimations of epistasis and QTL environment interactions. A number of softwares that are available for QTL interval mapping include MAPMARKER/QTL QTLSTAT, QTL Cartographer, MAPQTL, QGENE, QTLMapper, etc.

Marker Aided Selection (MAS) in Plant Breeding : If a molecular marker locus is tightly linked with a gene of interest, indirect selection for the gene can be exercised by selecting for a particular marker phenotype, which is linked with the desired phenotype for the economic trait. Such an indirect selection has several advantages including the following :

(i) During backcrossing, selection for a recessive gene can be exercised even in a heterozygous plant, if co-dominant makers like RFLPs are used; this will eliminate the need of selfing after each backcross as required in conventional backcrossing;

(ii) During backcrossing, indirect selection may also be desirable for dominant genes in some cases, because selection can be exercised at the seedling stage and the plants not carrying the desirable allele can be weeded out early, thus saving space and expense;

(iii) During pyramiding of genes (when it is desired to incorporate more than one genes for the same trait) for traits like disease or insect resistance, it is not possible to select for the presence of an additional gene in the presence of an existing gene for the same trait; under such a situation, molecular markers associated with different genes will differ and can be conveniently used to follow any number of genes during the breeding programme;

(iv) Selection for mapped QTLs representing several chromosome segments contributing to the quantitative trait, can be exercised simultaneously with the help of individual molecular markers associated with each of these chromosome segments.

Identification of Breeding Lines and Varieties using Markers: It has been demonstrated that even closely related individuals of some species (e.g. humans) can be distinguished using DNA fingerprinting. Molecular markers like SSRs and AFLPs, that are equivalent in resolution to minisatellites of human beings have now become available in plants, so that genotype specific RFLP/SSR/AFLP patterns have been detected and used for the identification of breeding lines and varieties and also for the characterization of plant genetic resources (PGRs).

Characterization of Genetic Resources and Estimation of Genetic Diversity using Markers : Characterization of genetic resources becomes prohibitive due to limitation of time and resources, particularly when large collections are being evaluated. The collection may also have duplications, which should be eliminated and a well characterized core collection should be prepared. Such core collection (5-10% of large collections) may be characterized using biochemical and molecular markers in addition to morphological and agronomic traits. These biochemical and molecular markers may also be used to assess genetic diversity within and among accessions and to monitor changes during maintenance and regeneration of genetic stocks. Work on DNA fingerprinting for characterization, conservation and maintenance of plant genetic resources is being undertaken in India at NFPTCR (National Facility for Plant Tissue Culture Repository), New Delhi and at CDFD (Centre for DNA Fingerprinting and Diagnostics), Hyderabad.

In the form of biochemical markers, isozymes have been used, but their use is limited due to lack of polymorphism and a small number of loci and alleles available for analysis. The use of DNA-based molecular markers for characterization overcomes this limitation, since they provide substantial polymorphism and they can be unlimited in number. The only disadvantage of using molecular markers for characterization of genetic resources is the cost of material and labour which may gradually go down. Developments of more efficient PCR based markers will make the use of molecular markers more convenient and more economic.

Molecular Cytogenetic Maps Using Cytogenetic Stocks : Through a study of linkage of molecular marker loci with morphological or biochemical (isozymes) markers having known chromosome location, it was possible to locate DNA-based molecular marker loci not only on specific chromosomes, but also on specific regions of these chromosomes. For this purpose, nullisomic-tetrasomic lines (in wheat), monosomics (in cotton, maize), trisomics (in barley and rice) and various available translocations (in cotton) or deletion stocks (in wheat), were utilized. This resulted in the preparation of molecular chromosome maps for all major crops.

The above technique also allowed mapping of duplicate loci, when more than one fragments homologous to a probe were detected on Southern blots or more than one PCR products obtained using same SSR primer pair. While in maize, 28.6% sequences detected duplicate loci, in tomato these duplicate loci were less than 2% of the loci tested. This gave new evidence about the presence of frequent duplicate chromosome segments in maize rather than in tomato. More recently, duplicate loci have also been detected using whole genome sequencing, as done in *Arabidopsis*.

Molecular chromosome maps are also being prepared for wheat, rye, barley, rice, etc. and the availability of aneuploids has greatly facilitated this work. For instance, in wheat the use of nullisomic-tetrasomic lines facilitated the location of marker loci on specific chromosomes and the use of telocentrics allowed their location on specific arms. This has thus allowed in wheat the identification of molecular markers for each of the chromosomes arms belonging to 21 different chromosomes. In this connection, under the name 'International Triticeae Mapping Initiative (ITMI)', international collaboration was initiated to prepare maps for the tribe Triticeae and each homoeologous group was assigned to one research group. Twelve such workshops to assess the progress made and to plan future work were organised during 1990-2002. Rapid progress in preparation of molecular maps of different species of the Triticeae was made through these efforts.

Alien addition lines in wheat have also been utilized for assigning genes to specific chromosomes of the alien species involved. For instance wheat-barley addition lines have been successfully utilized for assigning genes on specific barley chromosomes and wheat-rye addition lines were used for a similar purpose for assigning genes on specific rye chromosomes. Oat-maize addition lines were similarly used for assigning markers to individual maize chromosomes. Genome specific probes (isolated from DNA of *Aegilops squarrosa)* have also been isolated to identify the D genome chromosomes in wheat and several other polyploid species carrying the D genome e.g. *Ae. cylindrica,* CCDD; *Ae. crassa,* DDMM; *Ae. ventricosa,* DDUU.

Molecular Physical Maps

Physical maps of DNA fragments, chromosome segments, whole chromosomes or whole genomes are based on physical distances. These physical distances are measured in terms of relative physical lengths or in terms of kilobase pairs between genes or molecular markers or in terms of relative distances of these genes/markers from the centromeres. For instance, the distance of a marker from the centromere may be expressed as a fraction of the length of the whole arm taken as unity, so that the distance of a marker from centromere will be indicated as 0.1, if its distance from the centromere measures one-tenth of the whole arm. Physical maps may also be obtained in the form of restriction maps, where for one or more enzymes, the restriction sites are shown to be separated by distances expressed in terms of kilobase pairs. Whole genome sequences are also being used for preparing physical maps in case of *Arabidopsis* and rice genomes. These physical maps were prepared in the past using deletions, restriction digestions of DNA fragments followed by electrophoretic fractionation, or by *in situ* hybridization with molecular probes as done in case of wheat. But with the availability of vectors like yeast artificial chromosomes (YACs), bacterial artificial chromosomes (BACs) or binary bacterial artificial chromosomes (BIBACs), physical maps are now also being prepared with these clones, whose positions on the map are separated physically by distances in terms of kilobase pairs.

Physical Maps using Molecular Markers : In the past, physical maps of genes for a variety of morphological traits in crops line tomato and maize were prepared using conventional cytogenetic methods. For instance, in tomato, physical maps of known genes were prepared by CM Rick and GS Khush in late 1960s, through induction of deletions in the pollen that were used for pollinating the recessive genetic stocks, so that pseudodominance of a gene in the hybrid, and identification of deleted region in pachytene chromosomes allowed physical mapping of the gene in the deleted segment. Similarly, in barley, T Tsuchiya achieved physical mapping of some genes, using acrotrisomics and metatrisomics, and in maize, D Weber used B-A translocations and maize-oat translocations for physical mapping of some markers genes. However, for want of appropriate cytogenetic stocks, physical maps of molecular markers could not be prepared in all corps. Nevertheless, physical maps of molecular markers have been successfully constructed in bread wheat using delection stocks (constructed by BS Gill and TR Endo), and in barley using translocation stocks (constructed by G Kunzel and also by others).

Physical Maps in Bread Wheat using Delection Stocks : Several reports of physical maps involving molecular markers are available in bread wheat. This became possible due to the availability of a set of ~400 delection stocks that were developed by BS gill and TR Endo in USA. These delection stocks (mostly terminal deletions), approximately 400 in number, are spread over all the 21 chromosomes. A marker could be located in the deleted region of a chromosome by the absence of this marker in the corresponding deletion stock, even when the marker was not polymorphic. In this manner, more than 1000 molecular markers (including RFLPs, SSRs) have been physically mapped in bread wheat. Another 20,000 EST loci have also been physically mapped. These physical mappes and the related information are available on the internet (GrainGene site <http://wheat.pwe.usda.gov/ggpages/maps>).

Physical Maps in Barley using Translocations : For construction of physical maps in barley, translocation breakpoints in each of the seven chromosomes involving 120 translocations were used as physical landmarks. Since each chromosome was involved in several translocations, it could be divided in sub-regions on the basis of breakpoints, in a manner analogous to that used in the deletion stocks of bread wheat. STS primers from >300 genetically mapped RFLP markers were prepared, and were used with microdissected chromosomes carrying translocated segments as templates for PCR reactions. This allowed identification of markers flanking each translocation breakpoint, leading to the preparation of physical maps.

Physical Map in Sorghum : An integrated map of sorghum genome was also published in the year 2000. For this purpose BAC pools were screened with multilocus markers like AFLPs and MITEs.

Physical Maps using in Situ Hybridization (ISH) : Utilizing molecular probes derived from cDNA or gDNA (even if they were not derived from the same species i.e. heterologous probes), *in situ* hybridization could be successfully utilized for mapping genes with multiple copies on specific chromosomes. For instance cDNA copies of wheat rDNA were used for *in situ* hybridization to mitotic metaphase chromosomes of bread whcat cv. Chinese Spring. It was found that 90% of the ribosomal genes were located on chromosomes, 1B and 6B and the remaining repeat units located on 5D. This had suggested that the diploid genomes like A genome, which had two chromosomal sites for rRNA genes (Gerlach *et al.*, 1980) had undergone a change after incorporation into hexaploid *(6x)* bread wheat. Similarly, information was collected on physical location of the 5S rRNA loci in wheat, rye and barley. Using *in situ* hybridization (ISH), it was found that 5S rRNA loci are located distal to the secondary constrictions on chromosome 1B of wheat and 1R of rye and on non-nucleolar chromosome in barley. An additional 5S rRNA locus in rye was located on 5R.

Physical mapping of the 5S rRNA multigene families in bread wheat was also achieved by combining the technique of *in situ* hybridization (ISH) with election mapping (Fig.). Terminal deletion stocks produced by using monosomic addition of a gametocidal chromosome from *Aegilops cylindrica*, were utilized for detection and location of twelve 5S rRNA loci on short arms of chromosomes of homologous groups 1 (1 AS, 1 BS and 1 DS) and 5 (5 AS, 5 BS and 5 DS). Specific chromosome arms and regions of the arms associated with only five different loci were determined using ditelocentrics (DT) and deletion stocks. The position of ISH sites on a telocentric chromosome was determined as a fraction of the total telosome length from centromeres. The positions of five loci of 5S rRNA, which were designated as *5S-Rrna A 1, 5S-Rrna D 1, 5S-Rrna A2, 5S-Rrna* 52 and *B2-Rrna D2* were 0.77, 0.96, 0.76, 0.63, and 0.64, on arms of 1 AS, 1 DS, 5AS, 5 BS and 5 DS respectively. The 5S rRNA locus on 1 AS was discovered for the first time during this study.

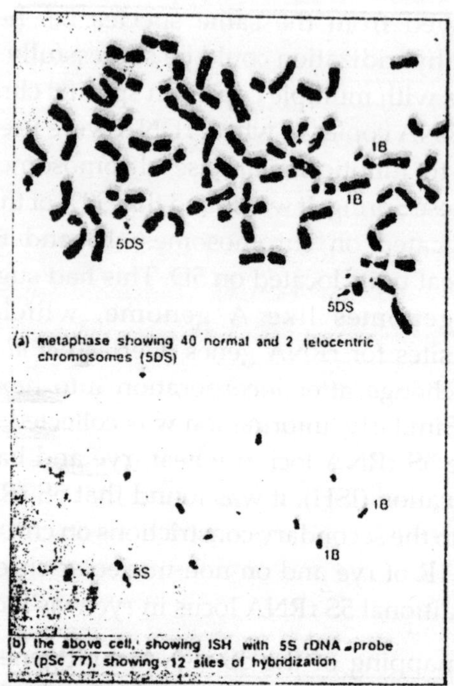

Terminal deletions for 5DS wheat chromosomes and ISH, used for detection of twelve 5S rRNA loci

In situ hybridization has also been used for physically locating the molecular markers earlier mapped in rice. It has been shown that distances on linkage maps (in cM) are not proportionate to physical distances on chromosomes.

Techniques have also been described, where each of two or three DNA sequences can be labelled with different non-radioactive haptens, each of which can be visualized with an independent detection system. It is, therefore, possible to detect two or three DNA sequences simultaneously on animals or plant chromosomes, in presence of each other, for physical mapping. For instance, in rye chromosomes, the location of pTa71 (rDNA sequence from wheat) and pSc 119.2 (a repeat sequence from rye) could be detected by labelling one with biotin and the other with digoxigenin and then using a mixture of the two for *in situ* hybridization.

In most of the above approaches involving ISH, individual DNA sequences (e.g. ribosomal RNA, or a specific repeat sequence) were physically localized and no real large physical mapping at the level of whole genome or a whole chromosome was undertaken. However, during 1995-2002, high resolution physical mapping of BACs anchored by genetic markers was undertaken using interphase and pachytene chromosomes. Fibre FISH technique was also used for physical mapping. The interphase and pachytene chromosomes being elongated give a better resolution in the physical map. Orientation of BACs with respect to telomeres and the centromere was also possible in pachytene chromosomes.

Genome-wide Physical Maps using Large-insert YACs, BACs and BIBACs : In all plants and animals, in order to facilitate genomics research, genome-wide physical maps are now being prepared using large-inserts in YAC, BAC and BIBAC vectors. These maps provide essential platforms for a variety of activities including the following : (i) large scale genome-sequencing, (ii) effective positional cloning, (iii) high throughput physical mapping of ESTs (expressed sequence tags), and (iv) target DNA-marker development. Several approaches that are available for the development of these large-insert physical maps include the following : (i) hybridization-based methods using interactive hybridizations, (ii) restriction-based fingerprinting methods and (iii) methods for development of integrated maps involving BAC end sequencing, fingerprinting and whole-genome sequencing. Of these three methods, the hybridization method suffers with the problem of the occurrence of repeat sequences, while integrated-map approach, based on sequencing and fingerprinting, is expensive. Therefore, the method of choice during 2001-2004 has been the restriction-based fingerprinting. In this restriction fingerprinting approach, the restricted fragments are end labelled and fractionated on agarose or denaturing polyacrylamide gels. Automatic sequencers were also used for electrophoresis-based fingerprinting leading to the preparation of large-insert BAC-based physical maps in *Arabidopsis thaliana,* an indica rice, and maize.

Integrated Physical Map of Arabidopsis Thaliana : During 1990-2000, several genome-wide physical maps of *Arabidopsis thaliana* genome were developed using large-insert YACs or BACs. These maps were also used for working out the whole genome sequences, which were published in December, 2000. The whole genome sequence, however, still had gaps in the heterochromatic and centromeric regions. In order to fill up these gaps, a whole-genome integrated physical map in *Arabidopsis* was prepared and publirhed towards the end of the year 2001.

Genomic-wide Physical Map in Indica Rice : In the year 2001, in an indica rice, a total of 21,087 BACs covering about seven times the rice genome were used for the preparation of a genome-wide physical map. The BAC fingerprints were obtained due to restriction digestion and fractionation on an automatic sequencer. The resulting fingerprints were scanned into image files, edited and created into FPC (fingerprint contig) database. Using FPC program, the overlapping clones were assembled into 585 contigs, each contig having 5-128 BACs (the contigs having less than 5 BACs were rejected). Two contigs were treated as overlapping, if the end clones of these contigs had 10 or more common bands. Using this criterion, extended contig's were developed and the number of contigs was reduced from 585 to 298, which covered 97% of the genome. The BACs of each contig were screened using DNA markers, and their correctness validated. In future, this physical map will certainly prove useful for genomics research in rice and other grass species.

Physical Map of Maize : Using inbred line B73, three BAC libraries (using three enzymes, HindIII, EcoRI and MboI) were prepared. Together the three libraries provided 27 fold genome coverage of the genome. The BAC clones of these libraries are being screened using markers earlier mapped using IBM population. The BAC clones are fingerprinted by digesting BAC clones with *HindIII,* and fractionating the fragments on agarose gel. The gel images are digitized with IMAGE software (Sanger Centre, UK). Contigs were assembled using Fingerprint Contig (FPC) software. In the first instance, 83,000 of the 450,000 BAC fingerprints were assembled into 13,000 contigs and 9,300 singletons (singletons are clones which have no overlap with any other BAC clone). A variety of markers including 90 RFLPs, ~1000 ESTs, a large number of AFLPs and MITES from the maize genome are being used to screen BAC libraries. A number of cDNA clones from sorghum were also used to assign markers to individual BACs. It is also proposed to develop at least 2000 SNPs from ESTs. SNPs will also be developed through direct sequencing of BACs and BAC ends. The resulting information will provide important anchor points for assembling individual BACs into contigs.

MaizeDB can also be accessed at http://www.agron.missouri.edu/, where maps, probes, primers, screening images and lab protocols are available. It also allows users to compare sequences of interest to all public maize sequences. For more information maize DB also provides database links including the following : (i) CUG 1 links to BAC contigs (ii) GenBank and ZmDB (*Zea mays* database at low State University) links for sequence and clone information.

Comparative Genomics and Collinearity/Synteny in Maps: Detailed molecular maps, available in all major crops, also allow comparative genetic analyses in several groups of plants, involving families like Poaceae, Solanaceae, Fabaceae and Brassicaceae. This will be briefly illustrated using the examples of the grass family, and those from Solanaceae. It has been shown that all genomes of grass species can now be described in terms of their relationships to a single reference, i.e. rice genome. Less than 30 linkage blocks are needed to represent genomes of all grass species including the following : wheat, rye, barley, oats, maize, sorghum, sugarcane, rice, wild rice (*Zizania palustris*), foxtail millet, pearl millet, finger millet (*Eleusine coracana*) and tef (*Eragrostis tef*). Similar comparative genomic analysis has also been undertaken in the family Solanaceae, where the maps of tomato, potato and pepper were compared suggesting that 18 homoeologous linkage blocks from tomato cover 98.0% of the pepper genome. An ancestral genome was also proposed, which through translocations, inversions, dissociations or associations of genomic regions differentiated into tomato, potato and pepper genomes.

The comparative analysis of molecular genetic maps also helped in predicting the location of genes controlling key agronomic traits, the only exception being the disease resistance genes, which undergo changes much faster. Two important examples of genes that have already been studied through comparative genome analysis include gibberilin-insensitive dwarfing genes which have been found to be orthologous between wheat and maize and the QTLs controlling shattering, which have been found to be orthologous among foxtail millet, maize, sorghum, rice and pearl millet.

Another important feature which has been brought to light through comparative genome analysis is the presence of large number of duplicate segments, suggesting that several species that are believed to be diploid may have an ancient polyploid origin (e.g. maize and *Arabidopsis*). For instance, large scale duplicate RFLPs were earlier detected in maize and more recently, 60% of the *Arabidopsis* genome has been shown to carry duplications suggesting tetraploid origin in both these cases.

Microcollinearity in DNA Sequences of Small Genomic Regions : Studies on molecular maps across related plant species, as above, though revealed significant conservation of gene content, gene order and gene homology, this comparative analysis had its own limitations. For instance, only up to about one marker per 10 centimorgan genetic distance is available for comparative analysis, thus making it difficult to analyse small deletions, duplications and inversions involving only a few centimorgans. Moreover, many genes have multiple homologs within the same genome thus making it difficult to establish orthology between genes from different but related species. In view of these limitations, in recent years, instead of markers, DNA sequences representing small regions of genomes have been used for comparative analysis, sometimes resolving what is described as microcollinearity. Although very few plant, genomic sequences have been used for such a study, some patterns have been delineated on the basis of available sequences. Comparative analysis of gene sequences and repetitive DNA sequences has shown that they evolve at different rates.

It has been shown that the gene sequences are more conserved, although disease genes have been shown to change rapidly. It has also been shown in several studies that the gene density differs and is related with genome size. For instance, at *Sh2/A 1* region, the intergenic regions were 7 times longer in maize than in the orthologous regions in rice and sorghum. Similarly in 280 kb region at the *adhl/u22* locus in maize, the genes were separated by more than 120 kb containing 10 retroelement families, but in sorghum the orthologous genes were only 50 kb apart. It has also been shown that microcollinearity may occur not only between related species but also between more distantly related species. For instance, in the *adh* region, collinearity is greater between closely related maize and sorghum than between either of these two species and the distantly related rice. In contrast to this, wheat shows greater collinearity with distantly related barley than with the closely related rye.

Molecular Transcript (EST) Maps : With the emphasis on functional genomics research, and with accumulation of 'expressed sequence tag' (EST) data for number of plant genome, efforts were also under way at the beginning of the present century to prepare transcript maps in crop plants. These maps could be genetic, cytogenetic or physical in nature. A composite map with 272 loci (spread over 1573 cM) of the expressed sequences and phenotypic traits in sunflower was published in 1999. The construction of this map involved anonymous cDNA markers. A rice transcript map was published in 2002 and barley transcript maps will be published in 2004-05. Ten different laboratories in USA are also busy in preparing a transcript map in wheat involving 10,000 unique ESTs.

Genomics for Evolutionary Studies : DNA sequences have also been utilized for evolutionary studies, where phylogenetic trees could be prepared on the basis of similarities between sequences of individual genes like 5S rRNA genes, 18S rDNA genes, *rbcL* and several other genes. Already by 1980s, large sequence data and sophisticated computer based analytical tools had become available for this purpose. However, in most of these earlier studies, sequences of one or two genes were utilized at a time, and the sequences of individual genes like 18S rDNA and *rbcL* gene did not find the most parsimonious trees. Therefore, in a study conducted in 1999, DNA sequences of plastid genes *rbcL* and *atpB* and the nuclear gene for 18S rRNA of 560 species of angiosperms and seven non-flowering plants were utilized together for preparing a well-resolved phylogenetic tree. Similarly, a study of DNA sequences of five mitochondrial, plastid and nuclear genes *(atpl, matR, atpB rbcL,* 18S rDNA) of 105 species belonging to 103 genera and 63 families suggested that *Amborella*, Nymphaeales and Illicales.-Trimeniaceae- *Austrobaileya* represent the first stage of Angiosperm evolution, with *Amborella* being sister to all other angiosperms. Almost the same conclusions were also drawn through a study of duplicate phytochrome genes, *PHYA* and *PHYC.*

Whole Genome Sequences

However, most higher plants have genomes bigger than that of the biggest eukaryotic genome sequenced till 1999, and were, therefore, difficult materials for whole genome sequencing: (Table). Despite this, in several higher plants including a Crucifer weed, thale cress (*Arabidopsis thaliana)* and crops like rice, sorghum and maize, significant progress in the analysis of whole genome was made in the last decade of the 20th century and in the early years of the 21 st century. For instance, DNA sequence of the whole genome involving all the five chromosomes of *Arabidopsis thaliana* was completed and published in December 2000.

Some Important Crops and their Genomes.

	Common name	Species	Genome size (bp)
I.	Brassicas		
	Thale cres	*Arabidopsis thaliana*	1.3×10^8
	Oilseed rape/canola	*Brassica napus*	1.3×10^9
II.	Cereals		
	Rice	*Oryza saliva*	4.3×10^8
	Barley	*Hordeum vulgare*	5.2×10^9
	Wheat	*Triticum aeslivum*	1.6×10^{10}
	Maize/corn	*Tea mays*	2.5×10^9
III.	Legumes		
	Garden pea	*Pisum sativum*	4.2×10^9
	Soybean	*Glycine max*	1.1×10^9
IV.	Solanaceae		
	Potato	*Solatium tuberosum*	1.8×10^9
	Tomato	*Lycopersicon esculentum*	1.0×10^9

Four drafts of the complete genome sequence of rice have also been completed, two of them by private companies (one by Monsanto in April, 2000 and the other by Syngenta in January, 2001) the third by BGI (China) and the fourth by IRGSP. Two of these draft sequences of rice genome (one each by syngenta and BGI) were published in Science (USA), on April 5, 2002. The complete genome of rice is now targeted to be fully sequenced (high quality sequence) by the year 2004-2005, and hopefully the genome of maize will also be sequenced eventually. The progress made in this exciting area of plant genomics and its implications on plant biotechnology and crop improvement programmes will be briefly discussed in this section.

Thale Cress (Arabidopsis Thaliana) Genome : *Arabidopsis thaliana*, an annual crucifer weed called 'thale cress' has been a favourite material for plant scientists and has often been compared with *Drosophila* of animal geneticists.

An adult thale cress (Arabidopsis thaliana) plant, whose genome was fully sequenced in December 2000.

It is commonly found growing in the paths and walls of many gardens and is widely distributed around the world in many different climates and habitats, ranging from Arctic to the equator. Its adaptation to diverse climatic conditions, therefore, provides unique opportunity to study adaptation and the molecular mechanism underlying this adaptation. It has a small adult plant size, small generation time (few weeks), each plant giving large number of offspring. It also has a small nuclear genome of about 130 Mb, comprising 5 chromosomes (2n = 10) and has less than 10% repetitive DNA in its genome. These attributes facilitated a community of scientists to utilize this weed in both classical laboratory experiments and molecular biology work leading to our understanding of a number of growth processes in higher plants. It also led to characterization of a number of genes that may be useful for improvement of our crop plants. This research activity of *Arabidopsis* community was later supported by an international collaboration (involving scientists from USA, Europe and Japan), called The *Arabidopsis* Genome Initiative (TAGI), which began sequencing the *Arabidopsis* genome in 1996. The sequencing of ~116 Mb of DNA comprising 26,000 genes and almost the entire genome of *Arabidopsis* was completed and published in 14 December, 2000 issue of Nature, a prestigious journal published from UK.

Sequencing Strategy used : For genome sequencing, strain Columbia of *Arabidopsis thaliana* was chosen. BAC/PAC genomic libraries were used to assemble a physical map, which was integrated with the genetic map. This was then used for assembling sets of contigs into sequence ready minimal tiling paths. Only about 116 Mb of the genome could be sequenced, the remaining about 10 Mb that remained unsequenced represented the centromeric and rDNA repeat regions.

Genome Sequences of Five Chromosomes : The nucleotide sequences of the DNA contained in *Arabidopsis* chromosomes 2 and 4 were published in December, 1999 and that of the remaining three chromosomes (chromosome 1, 3 and 5) were completed in December 2000, which marked the completion of the sequencing work for 115.4 million base pairs (Mb) of about 125 Mb of this higher plant. This was the first higher plant and the fourth eukaryotic genome sequenced then, the other three eukaryotic genomes being those of yeast *(Saccharomyces cerevisiae)*, a nematode worm *(Coenorhabditis elegans)* and the fruitly *(Drosophila melanogaster)*. Among the eukaryotic genomes sequenced by then, the sequence of *Arabidopsis* genome was also the one with fewest gaps in the gene rich regions. For the first time it also provided sequences of the substantial proportion of centromeric DNA of each of the five chromosomes. The number of genes predicted to be carried by the whole genome comprising the five chromosomes, was 25,498 in genomic DNA sequences that were only 115,409,949 base pairs (the whole genome being about 125 × 10^6 base pairs). Therefore, the number of genes was higher than those predicted for either the fruitfly (19,099) or those for *C. elegans* (13,601), the only two higher eukaryotes that were completely sequenced by the end of the year 2000.

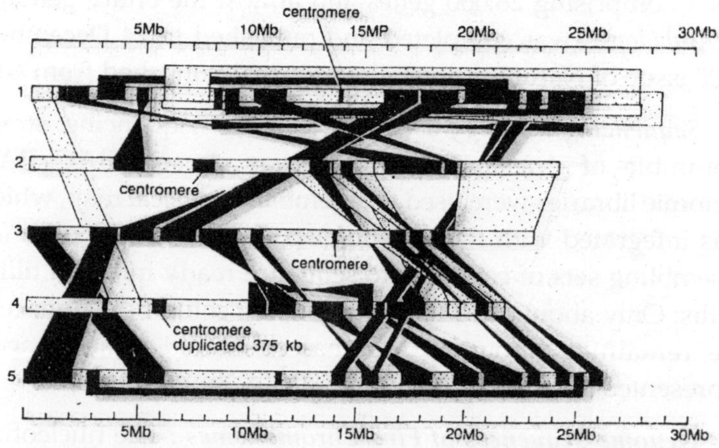

Five chromosomes (numbered 1 to 5) of Arabidopsis, showing duplicated regions, which are connected by bands; duplicated segments in reverse orientation are connected by twisted bands (redrawn from Nature).

***Duplications and Gene Redundancy*:** In *Arabidopsis* genome, most estimates suggest the presence of as many as about 26,000 genes (double the number present in *C. elegans*), although the functions of atleast 50% of these genes were still unknown, when the sequence was published in December, 2000. Surprisingly, 70% of the *Arabidopsis* genome is duplicated, the extensive duplications involving all the five chromosomes, thus reducing the number of unique genes that differ from each other to less than 15,000. Gene organization. *Arabidopsis* genes, in general are compact, each gene having many exons, each about 250 base pairs long and interrupted by short non-coding introns, 160-185 base pairs long (Table). These genes are about 4.6 kilobases apart, in contrast to much longer intergenic regions in many other bigger plant genomes. Small intergenic regions also suggest small regulatory sequences for the genes, which is also in sharp contrast to long regulatory sequences (upto 10 kb) known in many animal genes. The exons in *Arabidopsis* genes are GC rich (44%), relative to the GC content in introns (32%), which is considered to be a distinctive feature of plant genes.

Summary of Features of Arabidopsis Genome

Feature	*Chromosomes*					
	1	***2***	***3***	***4***	***5***	***Total***
I. Genome						
Length (Mb)	29.11	19.65	23.17	17.55	25.95	11.54
Number of genes	6,543	4,036	5,220	3,825	5,874	25,498
Gene density (per gene)	4.5	4.9	4.5	4.6	4.4	4.6
Gene size (bp)	2,078	1,949	1,925	2,138	1,974	1,997
Mean peptide length (bp)	446	421	424	448	429	434
Exons/gene	5.4	4.9	5.1	5.2	5.3	5.2
Exon size (bp)	247	259	250	256	242	251
Intron size (bp)	168	177	159	186	159	170
Number of ESTs ($\times 10^3$)	30.52	14.99	20.73	16.61	22.89	105.73
II. Proteome						
Number of Proteins	6,543	4,036	5,220	3,825	5,874	25,498

***Arabidopsis* Genes vs Animal Genes :** When the gene content of *Arabidopsis* was compared with those of the multicellular animal genomes sequenced then, it was found that in *Arabidopsis*, the number of singletons was about 11,500 and the number of distinct gene families was about 15,000, which was little higher than that predicted for nematode worm, *C. elegans* (13,601) and little less than that predicted for the fruitfly, *D. melanogaster* (18,424). However, many of the genes (presumably house keeping genes) were similar between plant and animals, suggesting common origin of animals and plants. There were also genes in *Arabidopsis* genome, which were unique to plants, like those involved in the synthesis and modification of water transporting channels and the cell walls. There are other genes in plants that encode for signal peptides directing transport of specific nuclear encoded proteins to chloroplasts, so that 25% of the genes in green plants like *Arabidopsis* encode signal peptides used in intracellular transport as against 5% in case of animals and fungi, where signal peptides are mainly required for intracellular transport to mitochondria, the chloroplasts being absent there.

***Arabidopsis* "2010 Project" :** During annotation of DNA sequences of *Arabidopsis* genome, it was noticed that the functions of about 50% of 26,000 genes of this genome were still unknown. Keeping this in view, in the year 2001, the National Science Foundation, USA announced funding of 28 new projects in the area of *Arabidopsis* functional genomics at an estimated cost of US $ 43.8 millions for a duration of four years. The project was designated "2010 Project" and was aimed to identify the functions of all genes present in *Arabidopsis* genome. A list of these projects awarded in 2001, along with abstracts of the projects are available at www.nsf.gov/bio/pubs/awards/2010fy01.htm.

***Rice (Oryza sativa) genome* :** Rice genome is 430 Mb in size, and is constituted into a haploid complement of 12 chromosomes. It is considered to be a model system for the study of cereal genomes, because on the basis of molecular maps, synteny conservation has been shown among all cereal genomes. The rice genome is also small in size, which is only three times the size of *Arabidopsis* genome, so that its sequencing was considered feasible and useful. The events that were involved in sequencing of the rice genome during 1990-2000 are listed in Table.

International Rice Genome Sequencing Project (IRGSP) : The work of sequencing the entire rice genome was initiated in Japan in the year 1998 with the establishment of an International Rice Genome Sequencing Project (IRGSP) led by Takuji Sasaki and was originally targeted to be completed by the year 2008. The work involving sequencing of each of the 12 rice chromosomes was allotted to individual laboratories around the world (including two laboratories in India, which sequenced a part of chromosome 11). However, before this international project could even take off, whole genome shotgun (WGS) strategy was used by two private companies to make the rough drafts of rice genome available in the years 2000 and 2001. In April, 2000, Monsanto with the help of LeRoy Hood at the University of Washington (USA) finished the rough draft of the rice genome, which had a 4x coverage.

Monsanto also announced that they will make this rough draft available to the IRGSP to help them expedite the completion of their project ahead of their targeted schedule. The IRGSP therefore, advanced its estimated completion by 5 years and announced that it will complete the sequencing of whole genome by the year 2003 instead of earlier target of 2008. Complete sequences of chromosomes 1 and 4 of rice were already published in November, 2002, and the sequence of chromosome 10 was published in early 2004. The sequences of other chromosomes should follow during late 2004.

A Summary of the Year-Wise Progress of Rice Genome Sequencing

Year	*Event*
1991	A 7 year rice genome project at a cost of 25 million was initiated in Japan Collinearity among cereal genomes established
1993	A 5-year rice genome mapping project was initiated at a cost 3.8 million in China
1997	An international consortium for sequencing of Nipponbare culture of japonica rice was formed
1998	IRGSP formally took shape and sequencing was targeted to be completed by 2008.
2000 (April)	Monsanto announced completion of a draft of rice genome sequence with four times coverage and promised to share the data with IRGSP and individual researchers
2000 (May)	Beijing Genomic Institute (BGI) announced that it plans to sequence the genome of indica rice using whole genome shotgun (WGS) within 2 years
2001 (January)	Syngenta announced sequencing of Nipponbare genome with six times coverage using WGS
2001 (October)	BGI completed sequencing of indica rice genome using WGS
2001 (December)	IRGSP decides to aim at 10 times coverage by the end of 2002
2002	Science published Syngenta and BGI draft sequences of japonica and indica rice genomes.

Two Draft Sequences of Rice Genome Published on April 5, 2002 : On April 5, 2002, two draft sequences of rice genome, one for each of the two major subspecies of rice *(japonica* and *indica;)* were published in American journal, Science. However, in both draft sequences, there are gaps to be filled and there may also be errors, which need to be rectified. The genome sequence for the *japonica* rice was worked out by a Switzerland based private company, Syngenta (Torrey Mesa Research

Institute or TMRI), so that it is proprietary in nature and has not been deposited in GenBank. However, this genome sequence will be available to scientific community on terms identical to those allowed for Celera's human genome sequence published on April 16, 2001 in the same journal. In contrast to this, the genomic sequence of *indica* rice has been worked out by Beijing Genomics Institute (BGI) and has been deposited at GenBank to be freely available to every body. It is believed that the above draft sequences of the rice genome will provide insights into the genomes of other cereals including maize, barley and wheat. IRGSP will also utilize these draft sequences to provide a much higher quality draft sequence of the Nipponbare cultivar of japonica rice by the end of 2004. The finished high quality rice genome sequence will also become available not before the end of the year 2004.

It should also be recognized that while Monsanto, Syngenta and BGI used the whole genome shotgun (WGS) sequencing approach to expedite genome sequencing, IRGSP is sequencing the rice genome by using 'clone by clone' approach or 'clone first-sequence later' approach of sequencing to provide high quality genome sequence with fewer gaps and fewer errors.

Japonica (left) and indica (right) rice plants, whose draft genome sequences were published in April, 2000.

Maize (Zea mays) Genome : The maize genome is almost as big in size as the human genome (~2.5 to 3.0 billion base pairs) and is highly repetitive with 65-80% of the genome consisting of retrotransposons, and another small fraction consisting of MITEs (miniature inverted transposable elements). Therefore, sequencing of this genome will be expensive and difficult. Nevertheless, a maize genome sequencing project is being initiated in USA. Since, maize genome differs from rice genome by as many as 15,000 rearrangements, it was believed that rice will be too distant to facilitate map based cloning in maize and other cereals like wheat, barley, sorghum, and oats.

The genome projects so far utilized either a minimal tiling approach or a whole genome shotgun (WGS) sequencing followed by computer assembly approach, the latter becoming difficult for maize genome due to its highly repetitive nature. None of these two approaches were considered suitable or desirable in case of maize genome. A third option, therefore, was considered desirable, which would involve identification, sequencing and assembly of all maize genes, which number approximately 50,000 and account for as little as 10-15% of the whole genome. Gene rich regions can be identified due to their hypomethylated nature, since only the gene poor regions, and the repetitive DNA get methylated; DNA with high *Cot* values and regions adjoining the transposons will also be rich in genes.

This strategy would involve sequencing of only the gene-rich regions and not the sequencing of large proportion of the genome involving the repetitive region, thus reducing the work substantially. It was also agreed that the inbred B73 should be utilized for genome sequencing, since it is already being used for developing a BAC physical map and the ESTs.

In the year 2001, data on complete sequence of about 150 maize genes and several BAC clones and the shotgun sequencing of about 5000 random maize genomic clones were already available. More BACs were being sequenced during 2001-04 under various genome projects that-were being funded at that time. A physical map was also prepared, which involved BAC clones, the corresponding contigs and the anchor markers; this map was considered essential for the maize genome sequencing project. Contigs as long as 1Mb (equivalents to 1-2 cM), each containing 20-40 genes became available in 2004. The complete genome (only the gene-rich regions) is estimated to be completed by the year 2005.

Other Plant Genomes

Other plant genomes that may be used for whole genome sequencing include sorghum (~750 Mb) among cereals and millets. *Lotus japonicus* (~450 Mb) and *Medicago truncatula* (~500 Mb) among legumes. *Brassica napus* (1300 Mb) among oilseeds and poplar (~530 Mb) among trees. Towards the end of 2001 and at the beginning of 2002, initiation of the first Legume Genome Project, i.e. sequencing of the entire genome of *Medicago truncatula* was announced in USA.

The Samuel Roberts Noble Foundation (founded by Lloyd Noble) announced that it will fund the project with an initial additional grant of US $ 5 million, thus raising the available grant for the project to US $ 10 million. The project will cost a total of US $ 45 million and is expected to be completed over a period of five years. The project will have two components: (i) sequencing the genome of *M. truncatula* and (ii) gene discovery and crop improvement, using the sequence information. The *Medicage* Genome Project will be undertaken by the Plant Biology Division of the Noble Foundation in partnership with Advanced Centre for Genome Technology (ACGT) headed by Bruce Roe, at the University of Oklahoma, which played a significant role in Human Genome Project and will also undertake the sequencing work in this Legume Genome Project.

Functional Genomics

Functional genomics involves determination of the functions of genes, and several innovative techniques are available to decipher the functions of these genes. Some of the approaches are specific to plant genomes and can not be used in animal or microbial systems. The approaches that are commonly used for deciphering the functions of the genes of higher plants are summarized in Table.

The different approaches of functional genomics, that can be used for the study of the functions of genes in plants include the following, which are shown in figure for rice.

Comparative Genomics

The genome sequences of *Arabidopsis* and rice that became available in the year 2000 and 2002 respectively, can be utilized for understanding the functions of genes in a variety of crops. Comparative genomic analysis can be undertaken at different levels of organization. Such studies at the chromosome level have been undertaken in grasses, using rice genome as the anchor genome. It was shown that 30 rice linkage blocks can be used to construct genome of each of a number of species. It was also shown that the 'gene space' occupies about 50% of the *Arabidopsis* genome and 24% of rice genome, the remaining genome being repetitive sequences and transposons.

Thus it was also clear that the absolute amount of gene sequence in different plant species is the same, irrespective of differences in genome size, which are attributed to proportion of repetitive sequences. Clustered organization of genes, with gene rich regions have also been identified in several crops. For instance, in wheat, barley and rice, one gene per 4-5 kb has been found in *Lrk* loci region, which is comparable to the gene density in *Arabidopsis*.

A summary or Methods used for Assigning Functions to Genes of Higher Plants.

Method	*Strategy*	*Criterion for assigning function*
BLAST search	Computer scanning of databases	Similarity with known gene
Synteny	Compare genetic maps from different species	Similarity in sequence and genetic map position
Chimeraplasty	Insert DNA/RNA hybrids into cells	Altered phenotype due to specific gene mutation
TUSC*	Insert DNA that jumps into genes	Generation of wholesale mutations in maize
Activation tagging	Insert DNA enhancers via bacteria	Generation of wholesale mutations in plants
RNA silencing	Infect tobacco with genetically altered TMV	Switching on or switching off individual genes
Microarray	DNA chips	Tracking activity using mRNA profile
Proteomics	2-D gels for protein expression	Tracking activity using protein profile
Metabolomics	Study of all metabolites	Tracking activity using metabolite expression

* TUSC = trait utility system for corn

whole-genome expression analysis : identify all genes expressed under select conditions in variety of interest

rice genome sequence

comparative genome analysis : identify candidate genes predicted to regulate trait of interest

reverse genetics to confirm prediction (overexpression and mutant analysis)

assess correlation of gene with trait in large populations

allele mining of diverse germplasm to identify useful variants

marker assisted breeding or genetic engineering of new crop varieties

whole-genome expression analysis : identify all proteins expressed under select conditions in variety of interest and determine interactions

Steps involved in the utilization of whole genome sequence for functional genomics leading to genetic engineering of crop plants.

There is also evidence for conservation of synteny (conserved clustering of genes/unique sequences) and collinearity (conserved order of genes or markers) among both related and highly divergent plant species like the following:

(i) rice and other grass species,

(ii) *Arabidopsis* and *Capsella rubella*,

(iii) *Arabidopsis* and *Brassica*, and

(iv) tomato and *Arabidopsis*.

While comparing genes, it has also been possible to distinguish between orthologous genes (derived due to speciation) and paralogous genes (derived due to gene duplication). The approach of micro-collinearity has also been used for study of gene function and for isolation of orthologous genes.

In Silico Discovery of Genes

Important computational methods for automatic identification of protein coding regions in genomic DNA sequences have been developed recently. This field of research is described as *in silico* discovery of genes and its use is increasing with the increase in cDNA and genomic sequences in databases. Improvements in this field of functional genomics involve development of new softwares and a better understanding of the mechanisms of transcription and translation. Several algorithms that are now available for annotation of available genomic sequences, have been successfully used for assigning functions to genomic sequences of *Arabidopsis* and rice.

Sequence Patterns in Gene : The annotation process, involving study of sequence pattern, includes identification of:

(i) start and stop codons of each gene;

(ii) regulatory sequences around each gene;

(iii) the position and size of the several introns in each gene (in *Arabidopsis*, on an average, each gene has 4-5 introns), and

(iv) sequences that are critical for transcription, RNA splicing and translation (promoter, branch point, polyadenylation site, ribosome binding site, etc.).

If all the above sequences are identified perfectly, then the protein coding region can be identified simply by removing the introns, concatenating all the exons, and reading off the protein sequence from start to stop. A number of gene finding softwares (GlimmerM, GeneScan, GeneMark.hmm) are available, which can be downloaded freely using the internet facility. However, for every species, new gene finders may generally be needed.

Sequence Similarity : A reliable method for identification of genes involves searching in genomic sequences of a particular species for a sequence, which is homologous to a sequence of known function in another species. For this purpose, genomic sequences, protein sequences and ESTs (expressed sequence tags) that are available in the databases are extensively used.

Such methods of gene finding by homology are often much more accurate than the *de novo* gene finding by computational methods, involving studies of sequence patterns (discussed above). However, all genes cannot be studied by homology search, since about 30% of genes in a species may be such, whose homologous sequences with known functions may not be available yet. In these cases, *de novo* gene finding by sequence patterns will still be useful.

Gene Finding in Arabidopsis and Rice : The method for annotation of genomic sequence, outlined above are already being used for *Arabidopsis,* for which whole genome has been sequenced and also in rice, for which at least four draft genome sequences became available by April 2002 and more are expected in future. However, it has been recognized that the gene finders that were made for *Arabidopsis* and used by '*Arabidopsis* Functional Genomics Consortium' (AFGC), do not work well in rice, so that new gene finders had to be designed for annotation of rice genome sequences.

Three gene finders that have been used for gene prediction in *Arabidopsis* include the following :

(i) GlimmerM,

(ii) GENESCAN +/- (a version of GENSCAN), trained for *Arabidopsis*, and

(iii) GeneMark.hmm.

These programs had already allowed identification of more than 1100 sequences, each containing only one gene (till early 2002). During 2000-2002, gene finders, specifically suitable for rice, were created. These included new rice versions of GlimmerM, GeneMark.hmm and FGENES (developed in 1995), which led to improvement in annotation process for rice genome. These gene finders may need to be retained, as more data on annotated genes accumulate for rice genome.

Efforts are also being made to combine the results of different gene finders listed above. Particularly, such an effort is useful, when different gene finders give different predictions. For instance, often a particular exon is correctly predicted by only one of the three gene finders. A combining algorithm was actually designed for *Arabidopsis* and was applied to the output of the three gene finders, leading to higher level of accuracy. Despite these improvements in gene finding algorithms in the year 2002, even the best systems predicted a complete gene correctly half of the time. Further improvement is expected during 2002-2010, when the number of genes with known functions grow and new algorithms are developed.

Management of Data for Functional Genomics : In recent years, there has been an exponential growth of data. For example, the number of sequences in GenBank increased from 4,864,570 in 1999 to 32,548,400 in 2004, which carried 38,000

Mbp. Some of the important databases and resources are listed in Table below. It may by noted that there are two main types of databases:

(i) single data-type databases which contain one type of information from many or all organisms and

(ii) multiple- data - type databases, which contain all types of information from a single organism (e.g. TAIR, MaizeDB, Soybase, etc.). The details of these databases, and the tools for their management are available elsewhere.

Selected Examples of Public Databases and Resources Used for Functional Genomics in Plants.

|I|	***Single organisms***
	TAtR *(Arabidopsis);* MaizeDB, ZmDB (maize): soybase (soybeans); AlfaGenes (alfalfa); BeanGeans *(Phaseolus* and *Vigna),* SorghumDB *(Sorghum bicolor).*
|II|	***Multiple organisms***
	GeneBank (all); SwissProt/ trEMBL (all); CropSeqDB (crop species); COGS (completely sequenced organisms); Protein Data Bank (all); DDBJ (all); EMBL (all): SolGcnes (solanaceous species); ARS Genome Resource *(Arabidopsis,* barley, *Brassica,* forage grasses, millet and rice); TIGR Gene Indices (many organisms; not all); GrameneDB (rice and other grasses); Grain Genes (wheat, rye, oat, barley and sugarcane).

***Gene Expression and Transcript Profiling*:** Gene expression studies through transcript profiling are an important area of research in the field of functional genomics. In contrast to gene finding methods, where one gene is studied at a time, expression profiles allow analysis of the whole genome at the same time. A number of methods are available for transcript profiling. These methods can be divided into two group :

(i) Direct analysis involving either the nucleotide sequencing of cDNAs or the analysis of fragments of different sizes;

(ii) Indirect analysis, involving nucleic acid hybridization of mRNA or cDNA fragments.

Direct Analysis : Direct analysis by nucleotide sequencing includes the following methods :

(i) large scale EST sequencing,

(ii) serial analysis of gene expression (SAGE);

(iii) massively parallel signature sequencing (MPSS).

The other methods involving direct analysis of the fragments of different sizes are also available and include the following techniques :

(i) differential display (DD);

(ii) selective amplification of fragments derived from cDNA (cDNA-AFLP).

Indirect Analysis : High throughput analysis of expression profiles involves nucleic acid hybridization utilizing DNA chips and microarrays. These profiles describe the gene expression patterns in different organs/tissues and in different growth environments. The *Arabidopsis* transcription profiles proved to be valuable in gene discovery and in the characterization of regulatory pathways. Similar studies in several crops including rice are being initiated in several laboratories around the world.

EST Contigs and Unigene Sets

Although extensive data on ESTs for a variety of plants are available in the databases, there seems to be considerable redundancy for many gene transcripts. Therefore, available ESTs have been assembled in contiguous overlapping clusters, which have been described as contigs. The ESTs which appear as singles and which can not be assembled in contigs are described as singletons. A combined set of contigs and singletons is described as a unigene set, which represents the minimum number of genes, although rarely same gene may be a part of two contigs, when an overlapping EST between these two contigs is missing in the database. These unigene sets have been constructed in many species including *Arabidopsis*, tomato, rice, sorghum, bread wheat.

Annotation of a unigene set is possible by matching the unigene sequences with either the annotated proteome of *Arabidopsis* (http:/ / *www.Arabidopsis.org)* using BLASTx or by matches against the GenBank protein database (http:/ / www.ncbi.nim.nih.gov / Database / index.html). This annolation allows classification of unigenes into role classes (r.c), of which the following four classes contain the largest number of unigenes; metabolism (r.c.l), transcription (r.c.4), cellular organization (r.c. 30) and celluiar communication/signal transduction (r.c. 10).

A comparison of annotated EST contigs in tomato with those of *Arabidopsis* (which are fairly diverse) also allowed identification of a set of > 1000 unigenes that were described as conserved ortholog set (COS). It is anticipated that this set will be conserved over very divergent species and will prove useful in comparative mapping/genomics studies over a wide spectrum of plant species.

Insertional Mutagenesis

Another approach for functional genomics involving identification and study of new genes is the use of insertion mutagenesis, which may involve one of the following two approaches :

(i) transposon mutagenesis and

(ii) T-DNA mutagenesis.

In both cases, a known DNA sequence is introduced, which may be randomly inserted at many places in the genome. This insertion of a foreign DNA segment (transposon or T-DNA) generally leads to loss of function mutations or 'gene knockouts' and only rarely leads to gain of function mutations. In each case, phenotype will be altered, suggesting the function of the gene involved in insertional mutagenesis. The insertion sequence (transposon or T-DNA) also allows isolation of the concerned gene which can then be studied. This approach has been extensively utilized for identification and isolation of new genes.

In order to utilize fully the above approach for functional genomics, it is necessary to saturate fully the genome with insertion mutations. These insertion mutations are being produced in a variety of plant species. In *Arabidopsis* alone, at least 15,000 such transposon insertion mutations are now available. At Wisconsin (USA), more than 60,000 *Arabidopsis*

T-DNA insertion lines were generated using T-DNA vector pD991. To share this resource with others, in 1999 a 'Knockout Facility' was established at the University of Wisconsin, as a part of the *Arabidopsis* Functional Genomics Consortium (AFGC). Researchers may send PCR primers for a gene for which they want a knockout. PCR products of the corresponding knockouts are then supplied to the researchers.

5

Transgenic Animals

Another important application of genetic engineering is the transfer of foreign genes into our livestock. In organisms like bacteria and other microbes, or even in higher plants, the uptake of genes by cells is often described by the term 'transformation'. However in animals, this term has been replaced by the term 'transfection', because the term 'transformation' in animal cell culture is used to describe phenotypic alteration of cells. The usage of the term 'transformation' for 'cell alteration' has been unfortunate and discontinuation of its usage for this purpose is suggested.

Transfection or gene transfer in animals may be carried out at the cellular level to get transfected cells, which may be used for a variety of purposes such as production of chemicals and pharmaceutical drugs. It may also be undertaken for basic studies involving study of structure and function of genes. Although many mammalian cell lines have been regularly utilized for these purposes, transfection has also been achieved successfully for the production of transgenic animals. The improvement in livestock through transgenesis has already led to the following encouraging results : (i) increased milk production in cattle, (ii) increased growth rate of livestock and fish, (iii) large scale production of valuable proteins in milk, urine and blood of livestock enabling the use of transgenic animals as 'bioreactors' for 'molecular farming; (iv) improvement of wool production through production of transgenic sheep. These encouraging results have made transfection and production of transgenic animals a fascinating thrust area of research. In this chapter, different methods of transfection and other aspects of transgenesis in animals will be discussed. The available examples of successful production of transgenic animals will also be discussed.

Gene Transfer or Transfection

In the past, the technique of somatic cell fusion has been used as a means for transfer of the whole genome from one organism to another. However, fusion can also be exercised between isolated nuclei, or between a nucleus and an enucleated whole cell. Similarly, the transfer of genetic material to whole cells can be exercised at the level of (i) individual chromosomes or fragments or (ii) isolated genes or gene fragments.

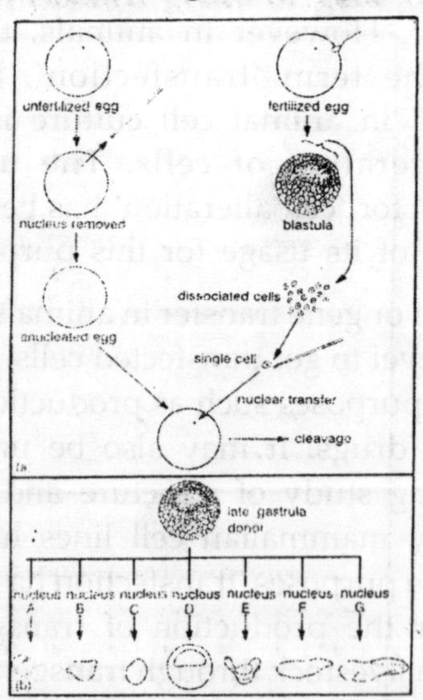

Fusion of karyoplasts with enucleated cells, in the presence of PEG (polyethylene glycol).

Transfection of Fertilized Eggs or Embryos : The production of transgenic animals requires transfection of specialized cells or embryos, since only eggs or embryos can develop into whole animals. Contrary to this, in plants any cell or protoplast may be transformed, since undifferentiated plant cells, being totipotent, can develop into whole plants through differentiation and growth. The eggs may be modified by transfer of whole nuclei, whole chromosomes or parts thereof or DNA fragments.

Transfer of Whole Nuclei (or Split Embryos) : Transfer of whole nucleus from a somatic cell of the superior donor to the enucleated egg can be achieved using the following steps (Fig.):

(i) enucleation of unfertilized egg is achieved by centrifuging cytochalasin-B-treated cells, such that the nuclei detach from the eggs and pellet at the bottom of the tube, leaving enucleated eggs in the supernatant;

(ii) karyoplasts (nuclei with only some residual plasma membrane) are similarly obtained from the blastula stage of the developing embryos of the donor;

(iii) karyoplasts derived from the donor are incubated with enucleated eggs in the presence of PEG (polyethylene glycol), and fusion is achieved;

(iv) the manipulated egg is transferred to the uterus of a surrogate mother for development.

Techniques have also been developed, where a superior developing embryo may be bisected into two parts using a surgical blade. Each bisected part of the embryo may be separately transferred to an enucleated unfertilized egg (Fig.). Such a manipulated egg may then be transferred to the uterus of a surrogate mother for further development.

Transfer of Whole Individual Chromosomes or Fragments: Chromosomes may be isolated from metaphase cells by hypotonic lysis. Incubation of these isolated chromosomes with whole cells after coprecipitation with calcium phosphate results in their incorporation into the nuclei. Chromosomes once isolated may also be subjected to fractionation using density centrifugation or flow cytometry and individual specific chromosome pairs may be inserted into recipient cells.

DNA Microinjection into the Egg : Transfer of individual cloned genes is generally achieved by inserting into a fertilized egg, a large number of copies of a recombinant plasmid carrying the gene of interest, using the technique of DNA microinjection (Fig.). The DNA is microinjected into a fertilized egg before the fusion of male and female nuclei.

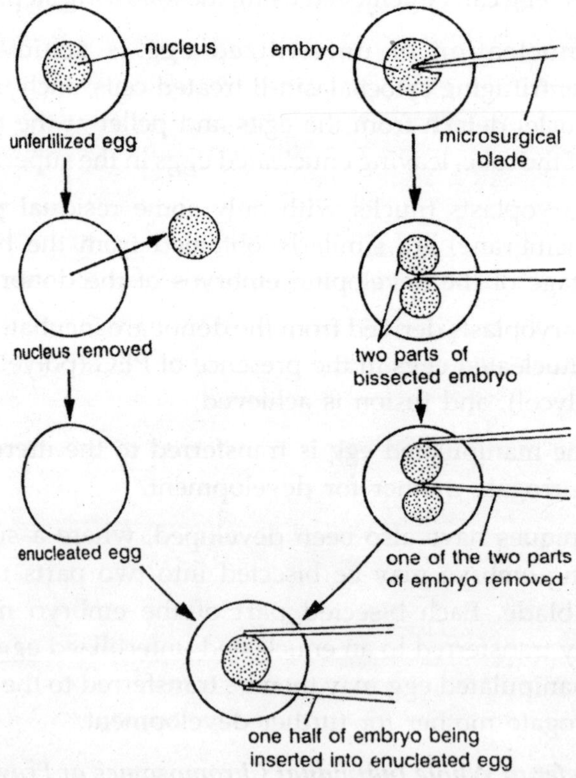

Splitting of an embryo, followed by transfer of one half of the embryo to an unfertilized egg whose contents were removed.

The egg is first immobilized by applying mild suction to the large, blunt holding pipette and DNA is then injected through the sharp end of a narrow glass microneedle. Through recombination at the DNA level, the inserted gene may get integrated into the host genome and inherited in a Mendelian manner. Transgenic mice have been produced in many laboratories, utilizing this technique. The method has also been successfully utilized for production of transgenic chicken, cows, fish, pigs, rabbits and sheep.

In *Drosophila,* the P elements (transposons) are used as vectors, so that a recombinant P element carrying the foreign gene is used for microinjection. Although the flies developing from microinjected embryos may not carry the inserted gene in all cells, some of their progeny may carry the gene in all the cells.

Virus Mediated Gene Transfer to Embryo : Retrovirus infection of embryos has also been used for the production of transgenic mice. This virus has been found to be an efficient vector system for animals (see later for details). The virus carrying the gene of interest transfers it into the genome of embryonic cells leading to production of transgenic animals.

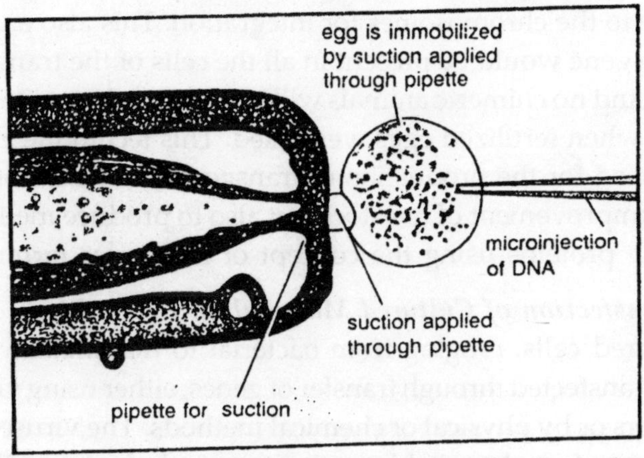

Microinjection of foreign DNA into a fertilized egg for production of transgenic animals.

Transfection of Unfertilized Eggs : The above gene transfer procedure involving transfection of either the fertilized eggs or the developing embryos is time-consuming and expensive. In this method, the efficiency of gene transfer leading to the production of transgenic animals was rather low. Only 1% to 10% of the animals that developed from fertilized eggs inoculated with a foreign gene had the gene integrated in its genome. Many of these animals were also mosaic, having transgene only in some of the cells, but not in others. Sometimes the transgene was present only in the somatic cells and not in the germ line, so that the gene was not transferred to the next generation. Consequently, the production cost for each transgenic animal could be as high as US $500,000. Keeping this in view and using retrovirus as a vector, efforts were made to introduce a foreign gene into bovine (cow) eggs before they were fertilized.

In 1998, this led to the production of transgenic cows, carrying a gene coding for hepatitis B surface antigen, with 100% efficiency. For this purpose, oocytes arrested at metaphase were isolated and used, so that the nuclear membrane was absent and the gene did not encounter any barrier during its journey to the chromosomes for integration. This also ensured that the gene would be present in all the cells of the transgenic animal and no chimeric animals will be produced, as witnessed in case when fertilized eggs were used. This technique is now being used for the production of transgenic animals, not only for the improvement of livestock but also to produce medically valuable proteins using the concept of molecular farming.

Transfection of Cultured Mammalian Cells : Many types of cultured cells, ranging from bacterial to mammalian cells, can be transfected through transfer of genes, either using viruses as vectors or by physical or chemical methods. The viruses that have been found stuitable as vectors include retroviruses, vaccinia virus, adeno-associated virus (AAV), herpes viruses and bovine papiloma virus. In mammalian cells, transfection frequency also increases dramatically, if DNA is co-precipitated with calcium phosphate in the recipient cells. In all these methods, some of the transferred DNA may integrate with the chromosomes of the cultured cells, thus resulting into stable transfection. If the donor DNA encodes a genetic trait, the recipient cells will express this trait.

Transfection of cultured mammalian cells has been utilized extensively for detection of cancer genes (oncogenes) and for gene therapy experiments. For detection and identification of a cancer gene, DNA from human tumour cell lines is purified, sheared into fragments (30-50 kilobase pairs in length), dissolved in phosphate buffer and then precipitated by the addition of calcium chloride. The solution is poured onto a layer of mouse 3T3 cells, which develop foci of transformed cells. These transformed cells can later be used for detection of cancer causing genes. (Mouse 3T3 cell lines are efficient for DNA uptake and have been used for decades to assay cancer causing genes).

Transfection of cultured cells has also been used for gene therapy, where the genes are inserted in cultured cells, which are then placed into the body of a patient for gene therapy. A variety of virus vectors are used for this purpose. Retroviruses were used in most of the early clinical trials (in 1990-91) including those on the two young girls suffering with adenosine deaminase (ADA) deficiency. Modified adenovirus and adeno-associated viruses are also being used. 'Naked DNA' is also used, for gene therapy, using, 'particle gun', 'liposomes' or 'electroporation device'. Future research will identify the safest and most efficient delivery methods for gene therapy.

Targeted Gene Transfer

In most of the methods for gene transfer described above, integration of the foreign gene occurs at apparently random chromosomal positions. Although even the 'misplaced' genes are subject to correct regulation under the control of regulatory sequences, random integration does not lead to precise replacement of defective gene at a homologous site. The gene transfer at homologous sites in the host genome is popularly described as targeted gene transfer. Initially, such targeted gene transfer by homologous recombination for replacement of a resident wild or mutant gene was possible only in bacteria and yeast. Later, for the first time in 1985, such targeted gene transfer was successfully achieved in human cells by Oliver Smithies at Madison (USA). The targeted gene transfer was possible due to the presence of homologous DNA sequences at the targeted site and in the vector carrying the foreign gene.

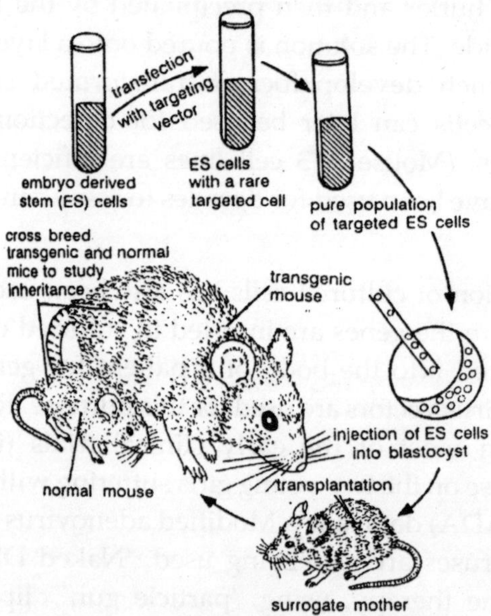

Transfection of stem cells with a vector carrying the gene of interest, followed by microinjection of transfected stem cells into a blastocyst leading to the production of transgenic mice.

Marker genes are also used for selecting cells transfected with transferred gene at the targeted site. Such a selection has been possible by several methods :

(i) by assaying the activity of an enzyme, as in hypoxanthine phosphoribosyl transferase (HPRT) enzyme;

(ii) by using a marker gene for antibiotic resistance that would enable cells to grow in a medium containing the antibiotic that would otherwise kill the cells;

(iii) by polymerase chain reaction (PCR), if sequence of the gene is available for designing PCR primers. Chimeric mice have actually been obtained by targeted gene transfer, where HPRT gene or homoeo-box containing genes have been modified in some embryonic cells. If these cells contribute to the production of egg or sperm, the transgenic mice will be produced. Success in targeted gene transfer has actually been achieved in many other cases also.

Triplex forming oligonucleotides (TFOs) and chimeraplasty using RNA-DNA hybrids have also been used for targeted gene transfer (consult author's another text book "Biotechnology and Genomics" for details). Gene targeting using embryonic stem (ES) cells. In this technique, embryo derived stem cells are cultured.

These stem cells are transfected with a vector containing the desirable gene. This allows targeting of gene to a particular site by homologous recombination. The rare targeted cells are isolated, multiplied and used for injection of expanded blastocysts, which are then implanted into surrogate mothers to allow the development to be completed. The developed animals can be checked for transgenesis. The different steps of this technique are illustrated in Figure.

Transgenic Animals

Transgenic Mice and other Animals : One of the first reports of transgenic animals published in December, 1982, involved transfer of growth hormone (GH) gene (from rat) fused to the promoter for the mouse metallothionein 1 (MT) gene. Since then many transgenic animals, including those in cattle, sheep, goats, pigs, rabbits, chickens and fish have been produced (Table) and will be utilized in future for a variety of purposes including:

(i) efficiency in utilizing food;

(ii) ability to give leaner meat;

(iii) ability to grow to marketable size sooner and

(iv) resistance to certain diseases.

More than this, efforts have also been made to use transgenic animals as living bioreactors. Transgenic animals produced for this purpose will secrete valuable recombinant proteins and Pharmaceuticals into their milk, blood and urine which can be used for extraction of these drugs. Several proteins including the following have already been obtained using transgenic mice :

(i) human alpha-1 antitrypsin (α -lAT);

(ii) human tissue plasminogen activator (t-PA);

(iii) human protein C(HPC);

(iv) human fibrinogen and

(v) a recombinant monoclonal antibody neutralizing TGEV (transmissible gastroenteritis coronavirus).

The level of synthesis often has been 1000 fold higher than what was observed in cell lines. This new possibility of manufacturing drugs through transgenic animals is often described as 'molecular farming' or 'molecular pharming'. Although, initially many experiments leading to the production of transgenic animals did not give commercially attractive results, success was later achieved in many cases. Some of these cases will be discussed.

A List of Transgenic Animals Produced and the Promoter, Enhancer and Structural Transgenes Involved in each case.

	Animal	*Genes transferred (promoter or enhancer/structural genes)*
1.	Mouse	*mMT/rGH, mMT/bGH, mMT/oGH, mMT/hGH, mMT/hGRF, mMT/hFIX*
2.	Chicken	*ALV, REV*
3.	Cow	*BPV, lactoferrin*
4.	Fish	*hGH, mMT/hGH, mMT/bGal, cd-crystallin SV/hygro, AFP*
5.	Pig	*mMT/hGH, mMT/bGH, hMT/pGH, MLV/rGH, bPRL/bGH*
6.	Rabbit	*mMT/hGH, hMT/hGH, rbEu/rbc-myc*
7.	Sheep	*mMT/hGH, mMT/TK, mMT/bGH, mMT/hGRF, oBLG/hFIX, oBLG/halAT, oMT/oGH*
8.	Goats	A variant of *tPA* gene *(LAtPA)*

Note:—The list of genes; promoter / or enhancer is separated from structural gene by a slash (/). The source of the gene is given by a lower case preceding the gene : b = bovine(ox): c = chicken : h = human : m = mouse ; o = ovine (sheep); p = porcine (pig); r = rat; rb s rabbit.

Gene abbreviations—*AFP* = antifreeze protein; *ALV* = avian leukosis virus : *alAT* = al anti-trypsin ; *BPV* = bovine papilloma virus; *Eu* = immunoglobulin heavy chain; *FIX* = factor IX: *GH* = growth hormone: *GRF* = growth releasing factor; (iGal = P galactosidase; *hygro* = hygromycin: *BLG* = (i-lactoglobulin: *MT* = metallothionein : *MLV* = moloney murine leukemia virus; *c-myc* = a cellular proto-oncogene; *REV* = reticuloendotheliosis: *PRL* = prolactin; *SV* = SV 40; *TK* =thymidine kinase: *tPA* = tissue-type plasminogen activator.

Transgenic Sheep

The rate of transgenesis in sheep is very low (0.1 to 0.2%). This can be improved, if only transgenic viable embryos (after necessary checking) are transferred to surrogate ewes (female sheep). Embryos at 8-16 cell stage can be split into two parts, one for continued culture and the other for detection of integrated genes using polymerase chain reaction (PCR). Although microinjection is the most common method for DNA delivery, gene targeting may be increasingly used in future. In this approach, embryonic stem (ES) cells in culture are transfected with a vector which targets the gene to a particular site by homologous recombination (as discussed above). This technique, though successfully used in mice, has yet to be applied to sheep, where ES cells will have to be isolated first.

The first reports of transgenic sheep were published by J.P. Simons (1988) of Edinburgh. Two transgenic ewes were produced, each carrying about 10 copies of human anti-hemophilic factor IX gene (cDNA) fused with the 10.5 Kb BLG gene (BLG = β -lactoglobulin). BLG gene was used, because it is necessary for specific expression of gene in mammary glands. Consequently, the gene had a tissue specific expression and ewes secreted human factor IX (or alpha-1 antitrypsin) into their milk(this human factor IX is active, even though the expression of transgene is low.

The transgenic ewes were born in early summer of 1986 and were successfully mated same year in December. In 1987, each ewe gave birth to a single lamb. Each lamb inherited *BLG-FIX* transgene and secreted factor IX in the milk. This programme of production of transgenic animals by J.P. Simons at Edinburgh was funded by Pharmaceutical Proteins Ltd. (Cambridge, UK), due to its commercial appeal.

In another report (published in 1991), also from Edinburgh, five transgenic sheep were produced (Alan Colman and colleagues). In all these cases, transgene involved fusion of the ovine β lactoglobulin gene promoter fused to the human α -1 antitrypsin (h α -lAT) gene (ovine = of sheep origin). Four of these animals were female, and one male. In one female the protein hoc-1 AT reached a level of 35 grams per litre of milk. The protein purified from milk had a biological activity indistinguishable from human plasma derived antitrypsin. The deficiency for ha-lAT leads to a lethal disease emphysema, which is a common hereditary disorder among Caucasian males of European descent. Therefore, any strategy giving high yield of this protein economically will be most welcome. In view of this, transgenic sheep with ha-lAT gene will prove very useful as a bioreactor.

Recombinant DNA technique can also be used to increase the ability of sheep for wool growth. For this purpose, genes essential for synthesis of some important amino acids found in keratin proteins of wool, have been cloned and introduced in embryos to produce transgenic sheep. For instance, genes *(cysE* and *cysM)* for two enzymes (serine acetyltransferase = SAT and O-acetylserine sulphydrylase = OAS), involved in cysteine biosynthesis, were isolated from bacteria and cloned in a vector. These genes were introduced in sheep cells, ultimately leading to the production of transgenic sheep, where these genes are expressed. Growth hormone (GH) genes have also been introduced and can be used to promote body weight. Other genes involved in wool production have also been cloned and will be used for transgenesis, thus increasing the potential of wool production through genetic engineering.

Transgenic Pigs

The efficiency of the production of transgenic pigs is still very low compared to that of the production of transgenic mice. In mice, 2.5% to 6% of the microinjected eggs developed into transgenic mice, but in pigs this frequency was as low as 0.6% even when as many as 7000 eggs were injected. This low frequency is due to failure of pregnancy, unless four fetuses are formed in the womb (pigs give birth to piglets in litters). Despite this low frequency, transgenic pigs carrying growth hormone (GH) gene from bovine (of 'ox' origin) or human, and sheep globin gene have been produced (by V.G. Purse) at Agriculture Research Service, Beltsville, USA.

The pigs carrying hGH gene showed different levels of expression and only 66% of these animals showed detectable levels of hGH and bGH in their plasma. The animals grew a little faster but did not become large; similarly pigs with sheep globin gene did not show any expression of the transgene for unknown reasons. In these transgenic pigs, however, a modest increase of 10-15% in daily weight and 16-18% in feed efficiency was observed, which are though lower to those in mice, but are comparable to those obtained due to daily injection with pig growth hormone. It was also observed that there was a marked reduction in the subcutaneous fat in some of these transgenic pigs suggesting the possibility of producing leaner meat with lower fat content. These results may have a significant impact on the 9.5 billion dollar annual pig industry in U.S.A.

It is also reported that a long term elevation of growth hormone was generally detrimental to health. The pigs had incidence of gastric ulcers, arthritis and several other diseases. Therefore, techniques will have to be developed to manipulate better the transgene expression by a variety of methods (e.g. changing genetic background or modifying the husbandry regiment. Transgenics pigs will also be used for xenotransplantation in humans, since size of pig organs matches that of human organs (consult from author's another text book "Biotechnology & Genomics)".

Transgenic Goats

Transgenic goats were successfully produced initially by groups headed by John McPherson and Karl Ebert both in USA. These transgenic goats expressed a heterologous protein (a variant of human tissue-type plasminogen activator = *LAtPA)* in their milk. This protein is used for dissolving blood clots i.e. for treatment of coronary thrombosis. A cDNA representing *LAtPA* was linked with either the murine whey acidic protein (WAP) promoter or a p casein promoter in an expression vector and used for injecting early embryos obtained surgically from the oviducts of superovulated dairy goats.

These injected embryos were either immediately transferred to the oviducts of recipient females (surrogate mothers) or cultured for 72 hours (blocked at 8-16 cells stage), before transfer to the uterus of recipient females. Of 29 offspring from 36 recipients, one male and one female contained the transgene. The transgenic female delivered five offspring, one of which was transgenic showing expression of *LAtPA* at a low level of few milligrams per litre of milk. In another case of a transgenic goat, few grams of *LAtPA* per litre of milk could be obtained. At this concentration, the dairy goat may become an economically viable bioreactor for human Pharmaceuticals.

Transgenic Cows

In most earlier attempts (in Canada) for the production of transgenic cows, embryos or fertilized oocytes produced *in vivo* were utilized. Fertilized oocytes or proembryos were surgically retrieved from superovulated and artificially inseminated cows. Microinjected zygotes were then transferred by surgery either directly into the oviduct of recipient cows or into temporary hosts like sheep or rabbits. In view of two surgical operations, this method is labour intensive and more expensive.

In 'The Netherland', in the year 1991, a technique was developed for *in vitro* embryo production. In this new procedure, oocytes obtained from the ovaries of slaughter house cows, were matured and fertilized *in vitro*. Their pronuclei were microinjected with a construct containing a bovine alpha-Sl-casein promoter (bovine = ox) driving a cDNA encoding the antibacterial human iron binding protein, 'lactoferrin'. The embryos were cultured to morula/blastula stage and then non-surgically transferred to recipient females. Two of the 19 calves born from 103 transferred zygotes were transgenic (one male and another female). Later in 1998, transgenic cows carrying a gene for hepatitis B surface antigen were produced through transfer of the gene to unfertilized oocytes arrested at metaphase (as discussed above). This procedure may facilitate the use of cows as bioreactors at the commercial level. During 1998-2001, transgenic cows for molecular farming have also been produced through cloning technique at the world's first clone farm in New Zealand.

Transgenic Fish

Attempts to produce transgenic fish started in 1985 and some encouraging results have been obtained. The genes that have been introduced by microinjection in fish include the following : (i) human or rat gene for growth hormone, (ii) chicken gene for delta-crystalline protein, (iii) *E. coli* gene for p-galactosidase, (iv) *E. coli* gene for neomycin resistance, (v) winter flounder gene for antifreeze protein (flounder = flat fish), (vi) rainbow trout gene for growth hormone.

The technique of microinjection has been successfully used to generate transgenic fish in many species such as common carp, catfish, goldfish, loach, medaka, salmon, *Tilapia*, rainbow trout and zebrafish. In other animals (e.g. mice, cows, pigs, sheep and rabbits), usually direct microinjection of cloned DNA into male pronuclei of fertilized eggs has proved very successful, but in most fish species studied so far, pronuclei can not be easily visualized (except in medaka), so that the DNA needs to be injected into the cytoplasm. Eggs and sperms from mature individuals are collected and placed into a separate dry container. Fertilization is initiated by adding water and sperm to eggs, with gentle stirring to facilitate the fertilization process. Egg shells are hardened in water. About 10^6 to 10^8 molecules of linearized DNA in a volume of 20 ml or less are microinjected into each egg (1-4 cells stage) within the first few hours after fertilization. Following microinjection, eggs are incubated in appropriate hatching trays and dead embryos are removed daily.

Since in fish, fertilization is external, *in vitro* culturing of embryos and their subsequent transfer into foster mothers (required in mammalian systems) is not required. Further, the injection into the cytoplasm in not as harmful as that into the nucleus, so that the survival rate in fish is much higher (35% to 80%).

Human growth hormone gene transferred to transgenic fish allowed growth that gave transgenic fish twice the size of their corresponding non-transgenic fish (goldfish, rainbow trout, salmon). Similarly, antifreeze protein (AFP) gene was transferred in several cases and its expression was studied in transgenic salmon. It was shown that the level of AFP gene expression is still too low to provide protection against freeze. There was also a report (in 1991) of the production of transgenic zebrafish from an Indian laboratory (Madurai Kamraj University). In this attempt, a plasmid containing rat growth-hormone gene was microinjected into fertilized zebrafish eggs, and its presence confirmed in adult fish.

6

Transgenic Plants

The significant use of transgenic plants is in the production of genetically modified crops. During 1990-2000, the progress in the production of transgenic plants to provide for GM crops has been so spectacular that by the end of 2002, transgenic crops tailored to market specifications were already being grown worldwide in an area of more than 58 million hectares (approximately 145 million acres). Transgenes will also be important in increasing the efficiency of crop production systems. For instance, transgenic plants resistant to herbicides, insects, viruses and a host of other stresses have already been produced. Transgenic plants have also been produced, which are suitable for food processing (e.g. bruise resistance and delayed ripening in tomato). For floriculture and forestry also, transgenic plants have been produced in several ornamentals and tree species.

Another exciting example is the production of male sterile (due to barnase gene) and fertility restorer (due to barstar gene) plants in *Brassica napus* and *Brassica juncea*, so that hybrid seed in future will be conveniently produced in these and other crops without manual emasculation and controlled pollination as practiced in maize. This has also eliminated the need for a search of cytoplasmic male sterility (cms) and fertility restoration system in crops, where hybrids are intended to be produced for higher yields. All these transgenic crops belong to the 'first generation' transgenic crops. The 'second generation transgenic crops' that were already being field tested at the end of the year 2002, will have enhanced nutritional quality, which is so important in the developing countries for millions of people who suffer from dietary deficiencies. 'Golden rice' that is rich in vitamin A and the rice fortified with iron are remarkable examples of these second generation transgenic crops. These achievements and prospects of the future use of GM crops will be briefly described in this chapter.

Despite the totipotency of plant cells, thus obviating the need of transforming specifically the germ line cells, as required in animals, the production of large number of transgenic plants till mid-1980s was restricted mainly to tobacco, petunia or tomato of the family Solanaceae.

A list of higher plants, where transgenic plants have been produced:

I. Herbaceous dicotyledons

1. *Nicotiana tabacum* (tobacco)
2. *N. plumbaginifolia* (wild tobacco)
3. *Petunia hybrida* (petunia)
4. *Lycopersicon esculentum* (tomato)
5. *Solanum tuberosum* (potato)
6. *Solanum melongena* (eggplant)
7. *Arabidopsis thaliana*
8. *Lactuca saliva* (lettuce)
9. *Apium graveolens* (celery)
10. *Helianthus annuus* (sunflower)

11. *Linum usitatissimum* (flax)
12. *Brassica napus* (oilseed rape: canola)
13. Brassica oleracea (cauliflower)
14. *Brassica oleracea* var. capitata (cabbage)
15. *Brassica rapa* (syn. *B. campestris*)
16. *Gossypium hirsutum* (cotton)
17. *Beta vulgaris* (sugarbeet)
18. *Glycine max* (soyabean)
19. *Pisum sativum* (pea)
20. *Medicago sativa* (alfalfa)
21. *M. varia*
22. *Lotus corniculatum* (lotus)
23. *Vigna aconitifolia*
24. *Cucumis sativus* (cucumber)
25. *Cucumis melo* (muskmelon; Centalonpe melon)
26. *Cichorium intybus* (chicory)
27. *Daucus carota* (carrot)
28. *Annoracia* sp. (horse radish)
29. *Glycorrhiza glabra* (licorice)
30. *Digitalis purpurea* (foxglove)
31. *Ipomoea batatas* (sweet potato)
32. *Manihot esculenta* (cassava)
33. *Ipomoea purpurea* (morning glory)
34. *Fragaria* sp. (strawberry)
35. *Actinidia* sp. (Kiwi)
36. *Carica papaya* (papaya)
37. *Vitis vinifera* (grape)
38. *Vaccinium macrocarpon* (cranberry)
39. *Dianthus caryophyllus* (carnation)
40. *Chrysanthemum* sp. (chrysanthemum)
41. Rosa hybrida (rose)

II. Woody dicotyledons

42. Populus sp. (poplar)
43. Malus sylvestris (apple)
44. *Pyrus communis* (pear)
45. *Azadirachta indica* (neem)
46. *Juglans regia* (walinut)

III. Monocotyledons

47. *Asparagus* sp. (asparagus)
48. *Dactylis glomerata* (orchard grass)
49. *Secale cercale* (rye)
50. *Oryza sativa* (rice)
51. *Triticum aestivum (wheat)*
52. *Tea mays* (corn)
53. *Avena sativa* (oats)
54. *Agrostis palustris* (turfgrass)
55. *Festuca arundinacea* (tall fescue)
56. *Musa acuminata* (banana)

IV. Gymnosperms (a conifer)

57. *Picea glauca* (white spruce)

Later reports of transgenic plants became available from other dicotyledonous families like Cruciferae, Leguminosae, etc. and from monocotyledonous families like Liliaceae and Gramineae. Some of these examples, where transgenic plants have been produced successfully are listed in Table.

Resistance Against Biotic Stresses

Biotic stresses have been responsible for serious losses in crop yields. The plant protection practices that have been used in the past to protect crops from these losses have reached a pollution level, which has became a major concern for environmentalists. Therefore, production of crops resistant to biotic stre~~s~~es including weeds, insects, viruses and other pathogens was the first priority in crop biotechnology. Significant benefits have already been derived from commercialization of these crops that are resistant to biotic stresses. Major benefits in USA alone are estimated to be billions of dollars every year.

Herbicide Resistant Transgenic Plants : Due to increasing concern about contamination of environment by herbicides, new herbicides are being developed that are safer and biodegradable. This has necessitated the development of resistance in crop plants against these new and safer herbicides. Herbicides, normally affect processes like photosynthesis or biosynthesis of essential amino acids (Table). Two approaches have, therefore, been used for the development of herbicide resistant plants: (i) In the first approach, we try that either the target molecules should become insensitive to herbicide or the target protein should be overproduced. (ii) In the second approach, a pathway is introduced that will detoxify the herbicide.

Modification of the Target : The target has been modified for developing resistance against atleast three herbicides (glyphosate, sulphonylureas and imidazolinones). The transgenic petunia plants resistant to glyphosate (active ingredient of Roundup herbicide) were developed by transfer of a gene for EPSPS (5-enol-pyruvyl-shikimat-3-phosphate synthase), that overproduces this enzyme. This overexpressing gene was isolated from plants selected for herbicide resistance. In some other cases, a gene *aroA* was isolated from the bacteria *Salmonella typhimurium* or *E. coli* and was transferred to tomato and/or tobacco. Another class of herbicides includes sulphonylurea compounds (active ingredients of Glean & Qust herbicide) and imidazolinones that inhibit the enzyme acetolactate synthase (ALS), which is involved in the bisoynthesis of branched chain amino acids like leucine, isoleucine and valine. Transgenic tobacco plants expressing a mutant ALS gene from tobacco or *Arabidopsis*, were produced

Mechanisms of Action of Different Herbicides and basis of Achieving Resistance Against them in Transgenic Plants.

Active principle of herbicide	*Inhibited pathway*	*Target product*	*Use*	*Basis of resistance*
I. Amino acid biosynthesis inhibitors				
1. Glyphosate (Roundup)	Aromatic amino acid biosynthesis	EPSPS	Broad spectrum	Overexpression of EPSPS gene; bacterial *aroA* gene
2. Sulphonylurea and Imidazolinones	Branched chain amino acids	ALS	Selected crops	Mutant ALS gene
3. Phosphinothricin (Basta)	Glutamine biosynthesis	GS	Broad spectrum	Gene amplification, *bar* gene : detoxification
II. Photosynthesis inhibitors				
4. Atrazine (Lasso)	Photosystem II	Q_B (32kDa protein)	Selected crops	Mutant *PsbA* gene. GST gene; detoxification
5. Bromoxynil (Buctril)	Photosynthesis	—	Selected crops	*bxn* gene: detoxification

that were tolerant to sulphonylurea herbicides. Similarly transgenic tobacco plants were produced by incorporation of genes for resistance against (i) L-phosphinothricin (PPT), which is the active ingredient of herbicide 'Basta' and inhibits glutamine synthase (GS), and (ii) atrazine which inhibits photosynthesis. These genes were isolated from *Medicago sativa* and *Amaranthus* hybrids respectively.

The transgenic tobacco plants expressed low level of resistance against the corresponding herbicides. Scientists at Plant Genetic Systems (PGS) and at the German Company, Hoechst, isolated a gene from *'Streptomyces hygroscopicus'*, which encodes an enzyme, capable of inactivating the herbicide 'Basta'. Transgenic plants with this gene have been produced and field tested, demonstrating effectiveness of this gene for protection against the herbicide 'Basta'.

Detoxification or Degradation of Herbicide : Detoxification or degradation of herbicides is the basis of selective effect of herbicide, so that the latter will kill the weeds and not the crop. A number of detoxifying enzymes have been identified in plants as well as in microbes. Some of these enzymes include (i) glutathione-S transferase or GST (in maize and other plants), which detoxifies the herbicide atrazine; (ii) nitrilase (coded by gene *bxn* in *Klebsiella pneumoniae),* which detoxifies the herbicide bromoxynil, and (iii) phosphinothricin acetyl transferase or PAT (coded by *bar* gene in *Streptomyces* spp.), which detoxifies the herbicide PPT (L-phosphinothricin).

Transgenic tomato plants using both *bxn* gene from *Klebsiella* and *bar* gene from *Streptomyces,* and transgenic plants in potato, oilseed rape *(Brassica napus)* and sugarbeet using only *bar* gene from *Streptomyces* have been obtained and were found to be herbicide resistant. Other target crops for engineered herbicide tolerance include soybean, cotton and corn. Herbicide resistant transgenic crops are already being commercially grown in large areas in several countries.

Complementary Systems of Insect Resistance in Crops.

System	Gene transfer	Patented	Comments
I. Substances occurring in plants			
1. Sugars (glucosinolates. polysaccharides)	No	No	The relative concentration of different sugars determines whether or not certain insects prefer a given plant or species (saccarose is preferred; not used in breeding).
2. Terpenoids	No	No	Large group of substances including pyrethroids, juvabione, pyrethrin I, sesquiterpenoid and phytoecdysteroids. Some terpenoids are common in cotton (i. e. gossypol) and used in classical breeding to increase insect resistance (restricted to red-flowering cotton oil; thus can not be used for feeding).
3. Alkaloids and glycoalkaloids	No	No	Includes nicotine, which has long been used as biological pesticide produced from tobacco extracts. Also used in classical breeding such as in potato (demissin and α -tomatin). Several Brassica species also contain sulfides (glycosyde) which are toxic to some insects and attract others.
4. Flavanoids	No	No	—

Contd...

System	*Gene transfer*	*Patented*	*Comments*
5. Phenols	No	No	Overexpression of polyphenoloxydase leading to increased production of quinonin which is toxic for insects. Unlikely to be used in genetic engineering due to complex chemistry and possible toxicity to mammals.
6. Protein antimetabolites			
6.1. Amino acids and primary storage proteins	No	No	All are secondary metabolites e.g. in wheat, gluten proteins are not digestible by some insects (*Eutygaster integriceps*) and thus confer partial resistance against insects. The presence of certain aminoacids generally increases resistance in rice.
6.2. Lectins inhibitors (also referred to as plant peptide hormones)	Yes	Yes	Lectins are common in the grains of cereals, particularly during germination. Certain lectins also have antifungal properties. Much work is going on with the snowdrop lectin (GNA); particularly promising for sucking insects (*Hotnoptera*), which cannot be controlled with *Bt*.
6.3. Protease inhibitors (includes trypsin and chymotrypsin inhibitors)	Yes	Yes	Act on exogenous proteolytic enzymes. Trypsin inhibitor has particularly broad insect spectrum and has been demonstrated to be synergistic with *Bt*; much work is being done with the cowpea protease inhibitor (CpTI).

Contd...

System	*Gene transfer*	*Patented*	*Comments*
6.4 α-amylase inhibitors	Yes	Yes	Widely occurring in seeds, particularly dicots. For gene transfer, the most effective so far has been the α-amylase inhibitor genes from the common bean.
II. Substance not occurring in plants			
7. *Btendotoxins*	Yes	Yes	See text for details
8. Secondary metabolites from bacteria	Yes	Yes	e.g. isopentenyl transferase (*ipt* gene) which affects the cytokinin biosynthesis in insects, leading to increased levels of toxins in insects. Such genes from bacteria have been transferred to several crops.
9. Other toxins	Yes	Yes	e.g. spider and wasp toxins have been transferred to plants for experimental purposes. Unlikely to be applied in crops due to effect of some toxins on mammals.
10. Smart proteins	Yes	Yes	Computer-aided design of novel proteins.

Serious losses in crop yields are caused due to insect pests. There are about 67,000 pest species that damage crops. Of these, 9,000 species are insects and mites, which are responsible for major yield losses in several of our important crops, particularly the tropical crops. Therefore, transfer of gene(s) providing resistance against insects in crop plants leading to the production of insect resistant transgenic crops has been a major application of biotechnology in agriculture sector. The insect resistant transgenic crops carrying Bt toxin genes have been shown to be very effective in controlling insect damage and are actually better than the currently used methods of insect control through insecticidal sprays. This not only allows saving in cost and time, but also reduces health risks and provides ecological benefits, since *Bt* toxins are highly specific against certain insects without affecting other specific insects (insecticides instead kill a broad spectrum of insects).

In classical plant breeding, following three systems for insect resistance are utilized:

(i) morphological barriers to insects (e.g. hairy leaves);

(ii) insect repellant or toxic substances released constitutively by the plant, or induced by insect damage, and

(iii) toxins that have either a repellant affect (e.g. quinonin) or dreadly effect (e.g. proteolytic enzymes such as trypsin inhibitor found in peas).

Several insect control strategies have been proposed in biotechnological approaches. These strategies are listed in Table, of which *Bt* is the most effective of all strategies used so far. A new insect resistance technology involving the use of smart proteins (computer designed) and VIPs (discovered in 1997) will certainly be utilized in future. Pyramiding of different genes active against the same insect will also become common in future. It is predicted that as many as five or more foreign genes against the same insect will be available in most insect resistant transgenic crops by the year 2007. Some of the novel systems include the following :

(i) control of insect growth through toxicity of their purine metabolic pathway;

(ii) use of *Beauveria virulent* with an active avaricide against *Lepidoptera;*

(iii) Use of 3-hydroxy-steroid oxidase against *Lepidoptera* and boll weevil;

(iv) use of a gene for carbonarin anti-insectant metabolites isolated from *Aspergillus carbonarius.*

However, the use of smart proteins, VIPs, engineered proteins (computer aided) and manipulation of metabolic pathways (e.g. production of azadirachtin, a toxic compound from neem tree) are believed to provide long term insect resistant transgenic plants. It has already been shown that metabolic pathways can be manipulated and engineered. It has recently been done successfully for insect resistance in maize, where a gene for limonene synthase has been used for accumulation of limonene, a toxic compound. Commercial application of this should be possible by the year 2005.

Bt and Biopesticides : *Bacillus thuringiensis,* commonly known as *Bt* is a gram positive soil bacterium. It is known for several decades that some strains of *Bt* kill certain insects (Lepidoptera, Coleoptera and Diptera), due to an insecticidal protein called δ - endotoxin, which disrupts the function of digestive system of these insects. It is also known that *Bt* protein is not harmful to mammals and is degraded within 20 seconds in digestive tract of a mammal, thus permitting safe handling and use of *Bt* formulations.

Different *Bt* strains are effective against different insects thus making it necessary to make a choice while using them as biopesticide. Although *Bt* was registered in USA in 1961, even today only <1% of all pesticides are based on *Bt*. The market for *Bt* biopesticides was US dollars 24 million in 1980 which grew to US dollars 107 million in 1989 and to US dollars 300 million in year 2000. There are about 100 *Bt* formulations in the market, all based on *Bt* protein, several containing as many as 5 different *Bt* toxins, and some of them containing recombinant proteins. Most of these formulations contain spores and crystalline inclusions that are released on lysis of *Bt* during growth. The potency of these formulations is 300 times that of synthetic pyrethroids commonly used and the toxin breaks down quickly in sunlight thus making it environment friendly.

Bt-based biopesticides also have the following disadvantages :

(i) it is relatively expensive;

(ii) agricultural machinery is needed for its application;

(iii) application need to be repeated several times in every crop season;

(iv) sunlight causes breakdown of the active ingredient;

(v) water washes the protein from the plant. Most of these difficulties, however, have been overcome in the production of transgenic crops.

Bt endotoxins and their genes. *Bt* toxins, symbolized as Cry meaning crystalline (reflecting the crystalline appearance of δ-endotoxin), were initially classified into four distinct classes, based on their host range. These are CryI (active against *Lepidpoptera),* CryII (active against *Lepidoptera* and *Diptera),* CryIII (active against *Coleoptera)* and CryIV (active against *Diptera)*. Many more classes have been added to these classes (e.g., CryV, CryVI and CryIX) and new strains have been shown to be active against nematodes and other pests. So far 50 genes for 50 Cry proteins are known, of which 28 genes were isolated from 14 different *Bt* subspecies that have been shown to be active against insects. Additional 30 Cry proteins described in the literature do not have any insecticidal properties; although they may be effective against nematodes and mites. These Cry genes are being variously modified to acquire new activities, thus extending the range of their insecticidal/nematicidal properties. The different *Bt* endotoxins (Cry) and their activity against specific insects are listed in Table.

Insect Resistant Plants Based on Bt : The above toxin genes *(bt2)* from *B. thuringiensis* have been isolated and used for *Agrobacterium* Ti- plasmid mediated transformation of a number of crop plants including tobacco, cotton and tomato plants. The transgenic plants were resistant to *Manducta sexta,* a pest of tobacco. Experiments of feeding the leaves of these plants to larvae of *M. Sexta,* showed 75%-100% mortality of the larvae, while the control plants carrying no transgenes were severely damaged. The presence of the gene *bt2* as well as that of the toxin protein synthesized under its control was also demonstrated by appropriate experiments. When inheritance of insect resistance was studied using crosses with normal control plants, F_1 showed resistance and F_2 generation exhibited expected segregation.

Since, a native gene often does not express the gene at high level, truncated versions of genes are synthesized with altered codon usage and elimination of certain sequences (codon usage means that for same amino acids, codons that are efficient in prokaryotes but not in eukaryotes are replaced by other degenerate codons known to be efficient in eukaryotes).

As many as 410 *Bt* related patents have been issued during 1985-96, 52% of them issued in the North America, 30% in Europe and Russian countries and the remaining 18% mainly in Japan. Further, about only 1/3rd of these patents were related to *Bt*- biopesticides and the remaining 2/3rd were related to genes or Bt-based transgenic crops. The major *Bt* genes include

CryIA(b), CryIA(c) and *CryIII(a). Bt-based* insect resistant transgenic crops, the genes employed, and the commercialization status of these crops are listed in Table.

Bt- crops as Substitute for Insecticide : World crop losses without the use of pesticides and other control measures is estimated to be 70% of the expected production. It is also estimated that, despite all efforts and measures, pests still destroy 50% of world production amounting to about US dollar 300 billion. In view of this, insect resistant crops offer great promise. Estimates of potential substitution value of Bt-transgenic plants for insecticide have been made and it is believed that out of a total cost of US dollar 8.11 billion spent on insecticide, US dollar 2.69 billion can be replaced by Bt-transgenic crops . This will include corn, cotton, rice and some fruits/vegetables.

The *Bt*-transgenic crops, while provide the above benefit, will also add expenses due to the cost of technology. However, in case of rice, International Rice Research Institute (IRRI) in the Philippines developed Bt-rice with a gene donated by Ciba of Switzerland (now Novartis). The germplasm will be made freely available to select national programmes and other interested parties. Similarly, at CIMMYT in Mexico, *CryIA(b)* and *CryIA(c)* are being used in corn. While these International Agricultural Research Centres (IARCs) may develop and make

Bt Endototoxin (Cry) and their Activity Against Specific Insect Species.

Cry protein[1]	*Origin (Bt subspecies)*	*Major target insects*	
		Order[2]	*Common names*
CryIA (a)	kurstaki	L	Silk worm, tobacco horn worm, European corn borer
CryIA(b)	berlineri	L&D	Tobacco horn worm, cabbage worm, mosquito

Contd....

Cry protein[1]	*Origin (Bt subspecies)*	*Order*[2]	*Major target insects* Common names
CryIA(c)	kurstaki	L	Tobacco budworm, cabbage hopper, cotton bollworm
CryIA(d)	aizawai	L	Several Lepidoptera
CryIA(e)	alesti	L	Tobacco budworm
Cry IB	thuringiensis	L	Cabbage worm
CryIB(c)	morrisoni	L	Several Lepidoptera
CryIC	entomocidus	L&D	Cotton leaf form, mosquito
CryIC(b)	galleriae	L	Beet army worm
CryID	aiawai	L	Beet army worm, tobacco horn worm
CryIE	keayae	L	Cotton leaf worm
CryIE(b)	aizawai	L	Several Lepidoptera
CryIF	aizawai	L	European corn borer, beet army worm
CryIG	galleriae	L	Greater wax moth
CryIIA	kurstaki	L&D	Gypsy moth, mosquito
CryIIB	kurstaki	L	Gypsy moth, cabbage hopper, tobacco horn worm
CryIIC	shanghai	L	Tobacco hom worm, gypsy moth
CryIIIA	san diego	C	Colorado potato beetle
CryIIIA(a)	tenebrionis	C	N/a
CryIIIB	N/a	C	Colorado potato beetle
CryIIIC	tolworthi	C	Colorado patato beetle
CryIIID	kurstaki	C	(*Aedes* and *Culex*)
CryIVA	israelensis	D	Mosquito (*Aedes* and *Culex*)
CryIVB	israelensis	D	Mosquito (*Aedes*)
CryIVC	N/a	D	Mosquito (*Culex*)
CryIVD	N/a	D	Mosquito (*Aedes* and *Culex*)
CryV	N/a	L&C	European corn borer, spotted cucumber beetle
CryDC	galleriae	L	Greater wax moth

N/a = not available; [1] *L=Lepidoptera; -D=Diptera; C=Coleoptera*

Note : 1 = I

Information About Cry Genes Used for Production of Insect Resistant Transgenic Crops

Transgenic Crop	*Transgene (s)*	*Status*
Potato	*CrylA(b), Cry IIIA(a)*	F.t, com.
Corn	*CrylA(b) CrylA (c), CBI-Bt*	Com.
Cotton	*Cry lA(a) CrylA (b), CryIA (c)*	F.t, com.
Tobacco	*CryIA(b), CryIA(c)*	Com
Tomato	*CryIA(b),*	F.t
Canola/repeased	*CryIA(b)*	F.t

CBI-Bt = Confidential Business Information
Γt. = Hcld trials (small scale, large scale)
Com = Commercialized in 1995/1996.

available the transgenic crops in the developing world, for several other crops being developed by private sector, the fanners will have to pay for the investments made, so that the seed of transgenic crops will have to be purchased at a premium. However, it is believed that the seed of transgenic varieties may be sold at prices identical to those of non-transformed variety as a means of boosting the market for transgenic crop varieties. In USA, farmers paid US $32/acre in royalties for Bt-cotton for an estimated saving of US $60-120/acre generally spent on insecticide. According to Monsanto (which marketed Bt-cotton), *this gives a saving of at least US $33/acre. Similarly, in case of Nature Mark™* potatoes, the premium per acre is US $16 to $72, compared to chemical control cost of US $30 to $120 per acre, a net saving of at least $14.00 per acre.

Management Strategies of Bt Deployment : One of the disadvantages of the development of insect resistant crops is the emergence of insect strains that may possess resistance against the toxin. This needs management strategies so that the resistance in the Bt-crops is durable. These strategies may be at the level of genes (pyramiding of genes or chimeric genes), at the level of gene promoters (tissue specific or wound inducible promoters), at the level of gene expression (low dose or mixtures) or at the level of field (mixture of genes, gene rotation, mosaic planting, refuge spatial or temporal).

Considerable success has been achieved in pyramiding of genes and in monitoring the expression of genes. But in every strategy, the refuge approach is considered necessary to avoid development of resistance in the insects. This is achieved by ensuring that there are always plenty of susceptible insects nearby for the few resistant ones to mate with, so that the population of resistant insects does not increase (susceptibility being dominant). Refuge areas are provided in two possible ways :

(i) only part of the field is planted with *Bt-crop*, the remaining part planted with conventional crop treated with insecticide;

(ii) the whole field is planted with Bt-crop and a small area nearby is planted with conventional crop, totally untreated.

In both cases, non-transgenic fields will generate susceptible insects in large number to mate with resistant ones developed in the transgenic field, so that resistant insects will not be able to mate with resistant ones leading to establishment of resistant strain in the population. Mixtures of transgenic resistant and non-transgenic lines in the same field have also been suggested and this approach is described as mosaics. (For more details, consult author's book *'Biotechnology and Genomics'*.

Genes for Protease Inhibitors : *(a) Gene for cowpea trypsin inhibitor or CpTI.* In cowpea *(Vigna unguiculata)*, trypsin inhibitor (CpTI) level was shown to be responsible for its resistance to attack by the major storage pest of its seeds (i.e. bruchid beetle = *Callosobrunchus maculatus).* CpTI was later shown to be toxic to a variety of insects (Table) but cowpea seeds with high level of CpTI are not toxic to humans. It was, therefore, considered desirable to transfer gene(s) for CpTI for production of transgenic insect resistant plants. A number of binary vectors were developed using Ti plasmid, where CpTI gene was joined with CaMV35S promoter, and one or more marker genes. The vector was mobilised into *Agrobacterium*, which was used to

Insecticidal Efficacy of CpTI in Artificial Diets and in Transgenic Plants (after Gatehouse et al., 1991).

Insects killed by CpTI in artificial diets	*Insects, against whom resistance noticed in transgenic tobacco plants (with CpTI)*
I. Lepidoptera	
Heliothis virescens	*Helothis virescens*
H. zea	*H. tea*
Spodoptera littoralis	*Spodoptera littoralis*
Chilo partellus	*Manducta sexta*
Autographa gamma	
II. Coleoptera	
Callosobruchus maculatus *Anthonomus grandis* *Diabrotica undecimpunctata* *Tribolium confusum* *Costelytra zealandica*	these insects do not attack control tobacco plants used as control
III. Orthoptera	
Locusta migratoria	

infect tobacco leaf discs, which led to the production of transgenic tobacco plants expressing high level of CpTi (as shown by western blotting) imparting resistance against a variety of insects. The CpTI gene in transgenic plants is stably inherited and there is no serious 'yield penalty'. Thus like Bt toxin, CpTI can also be used as a protectant against insect attack in transgenic plants. However, extensive field trials will be necessary before these transgenic plants can be released to the farmers for cultivation.

(a) *Gene for Alpha-amylase Inhibitor* (α *AI-Pv)* : The insect pest resistant transgenic plants produced using *Bt* gene, protect them from insects mainly in the field, but not during storage. The first transgenic plants, whose seeds, during storage, were resistant to pests (bruchid beetles), were produced in the year 1994 in pea. This was achieved by inserting a gene for α -amylase inhibitor (α AI-Pv), isolated from common bean *(Phaseolus vulgaris)*, and driven by a strong seed-specific promoter (from phytohaemagglutinin gene of bean; phytohaemagglutinin is a lectin). These transgenic plants produced seeds that contained 1.0-1.2% α AI-Pv and were resistant to cowpea weavils and Azuki bean weavils.

(b) *Genes for phytocystatins, cysteine proteinase inhibitors [e.g. oryzacystatin (OC) & corn cystatin (CC)]* : Phytocystatins which are cysteine proteinase inhibitors occur in a variety of higher plants, such as rice and corn. They probably have dual function :

(i) They regulate the action of endogenous cysteine proteinases in seeds during their ripening; for instance, in rice oryzacystatin (OC) inhibits oryzains and in corn, corn cystatin (CC) inhibits corn proteinases.

(ii) They recognize and inhibit exogenous proteinases, such as digestive enzymes in the gut of insect pests (cysteine proteinases occur in the guts of many insects and are necessary for the dietary life processes of these insects). Therefore the development of these insect pests can be inhibited by these cysteine proteinase inhibitors, which are also called phytocystatins.

Oryzacystatin (OC) occurs in rice seeds in concentrations (0.001-0.002%), which are insufficient for effective protection against insect pests. Moreover, it is quite deficient in the leaves, which are generally eaten by the insects. Therefore, for imparting resistance against insects, transgenic plants need to be developed, which have higher level of expression of these phytocystatins (eg. OC or CC). Attempts to produce rice plants with overexpression of OC were not successful. However, in rice plants, high level of expression (both in seed and leaves) of 'corn cystatin' or CC (~ 2% of, total soluble protein) could be successfully achieved. They offer an additional advantage, since CC has a wide inhibitory spectrum against various cysteine proteinases.

Genes for other Insecticidal Secondary Metabolites : Secondary metabolites produced by plants have also been implicated in resistance to insect attack (Table). However, biosynthesis of each of these metabolites involves a series of steps (sometimes even more than one biosynthetic pathway may be involved), each controlled by a separate gene. Furthermore, these genes are tissue specific in expression. These features make the production of transgenic plants difficult in this case. Protease inhibitors discussed in the previous section are also secondary metabolites, but their transfer is easier, since each is single gene controlled.

The production of transgenic plants by transfer of genes for entire multi-enzyme biosynthetic pathway (or for its augmentation) has now been possible (see later for vitamin A). For this purpose not only the transfer of genes is required, but the gene expression needs to be regulated, otherwise it leads to 'yield penalty' and also to toxic effect when consumed by humans and livestock. Defined promoter sequences are now available, which will, allow in the transgenic plants, a control over temporal and spatial expression of genes for biosynthetic pathways of secondary metabolites. Utilizing these facilities, transgenic plants with insect resistance due to secondary metabolites will be available in future.

Resistance Against Viral Infection

Cross Protection : It has been shown that if susceptible strain of a crop is inoculated with a mild strain of a virus, then the susceptible strain develops resistance against more virulent strains. This phenomenon is known as cross protection and has been used to reduce yield losses in crops like tomato against tomatc mosaic virus (TMV) in potato against potato spindle tuber viroid and in citrus against citrus tristeza virus. In most cases of cross protection, the symptoms of infection are delayed, and even the replication of virus is suppressed, although eventually the severe strain may be able to overcome the protection.

Examples of Secondary Compounds from Legume Seeds with Demonstrated Insecticidal Properties (after Gatehouse et at, 1991).

Compounds	*Bruchid (insect)*	*Legume*
I. Non-protein antimetabolites		
1. Alkaloids		
(i) 2, 5-dihydroxymethyl-3,4-dihydroxypyrrolidine (DMDP)	*Callosobruchus maculatus*	*Lonchocarpus* sp.
(ii) Castanospermine	*Callosobruchus maculatus*	*Castanospermum australe*
2. Non-protein amino acids		
p-Aminophenylalanine	*Collosobruchus maculantus*	*Vigna* sp.
Zabrotes subfasciatus		
3. Rotenoids	*Callosobruchus maculatus*	*Lonchocarpus salvadorensis*
4. Saponins	*Callosobruchus chinensis*	*Phaseolus vlugaris*
	Callosobruchus chinensis	*Glycine max*
5. Polysaccharides		
(i) Pectosans	*Callosobruchus chinensis*	*Phaseolus vulgaris*
(ii) Heteropolysaccharide	*Callosobruchus chinensis*	*Phaseolus vulgaris*
Acanthoscelides obtectus	*Phaseolus vulgaris*	
(iii) Polysaccharide (?)	*Callosobruchus chinensis*	*Vigna* spp.
II. Protein antimetabolites		
1. Lectins (phytohaemagglutinins)	*Callosobruchus maculatus*	*Phaseolus vulgaris*
2. α -Amylase inhibitors	*Zabrotes subfasciatus*	*Phaseolus vulgaris*
	Callosobruchus chinensis	*Phaseolus vulgaris*
3. Protease-inhibitors	*Callosobruchus maculatus*	*Vigna unguiculata*
4. Arcelin / LLP	*Zabrotes subfasciatus*	*Phaseolus vulgaris*
	Zabrotes subfasciatus	*Phaseolus vulgaris*

There are various disadvantages of this practice of cross protection such as (i) possibility of mutation in inducing mild virus strain; (ii) possibility of synergism between inducing virus and another unrelated virus; (iii) possibility of unnecessary spread of mild virus causing threat for future yield losses and (iv) possibility of some yield losses due to mild strain also. These disadvantages can be overcome, if a single gene imparting this benefit is transferred and transgenic plants are produced.

Transgenic plants have been produced in tobacco, tomato, and potato, which are resistant to a broad spectrum of plant viruses. Following six kinds of genes have been used for this purpose.

Gene for Virus Coat or Capsid Protein (CP) from Positive Strand RNA Viruses : Coat protein gene from tobacco mosaic virus (TMV), classified as a positive strand RNA virus, has been transferred to tobacco, and in the transgenic plant, expression of coat protein (CP) was observed. Further, when inoculated with TMV, the infection in the transgenic plants was very low and delayed relative to control plants that were not transformed.

This provides a new approach for producing virus resistant plants complementing the efforts of classical plant breeding in producing disease resistant crop varieties. However, in such transgenic plants, coat protein gene should be constitutively expressed and may thus have effects on the nutritional value of plants. Subsequently, this approach has been applied to a range of crops (tomato, alfalfa, tobacco, potato, melons, rice) for developing resistance against a broad spectrum of positive RNA strand plant viruses (e.g. alfalfa mosaic virus, potato virus X = PVX, potato virus Y = PVY and potato leaf roll virus). Potatoes have been produced which have coat protein genes and are tolerant to both PVX and PVY. DNA coding for a component of TMV replicase enzyme, was also transferred to tobacco plants, conferring resistance to TMV.

Genes for Resistance against Geminiviruses : Geminiviruses are single stranded DNA viruses causing important diseases in cassava, maize and other cereals. Coat protein in these viruses is not necessary for infection, so that the coat protein genes do not provide protection against these viruses. There are at least three examples of transgenic plants produced to provide resistance against individual geminiviruses:

(i) Tobacco plants *(Nicotiana benthamiana)* transformed with *AC2* gene (necessary for viral replication) of tomato golden mosaic virusc (TGMV), in the antisense orientation, showed good level of resistance against TGMV infection;

(ii) Tobacco plants transformed with a tandem repeat of subgenomic DNA B of African cassava mosaic virus (ACMV) provided protection against ACMV attack;

(iii) Capsid protein gene (VI) from tomato yellow leaf curl virus (TYLCV) when introduced in susceptible tomato hybrid plants *(Lycopersicon esculentum x L. pennellii)*, provided enhanced resistance, so that the transformed plants recovered from disease more effectively than the untransformed plants.

Gene for nucleocapsid : (N) protein from tomato spotted willt virus (TSWV). In earlier reports of resistance against viruses in transgenic plants, a number of positive (+) strand RNA viruses were involved. No report of resistance against any negative (-) strand virus was available. An important negative (-) strand virus is tomato spotted wilt virus (TSWV), which causes considerable yield losses in crops like tomato, tobacco, lettuce, groundnut, pepper and in ornamentals like *Impatiens, Ageratum* and *Chrysanthemum.* In this virus, genomic RNA is tightly associated with nucleocapsid (N) protein. This protein helps in (i) wrapping of viral RNA and also in (ii) regulation of transcription to replication switch during infection cycle.

For producing transgenic tobacco plants, N gene was associated with CaMV35S promoter and a leader sequence (to enhance expression). The transgenic plants showed expression of N-gene and also significant resistance to TSWV. A correlation was also observed between the amount of N protein and the level of resistance. The mechanism, involved in imparting resistance against negative strand RNA virus like TSWV, may differ from that involved in positive strand RNA virus, where coat protein (CP) gene has been used.

Satellite RNA and its use for I Transformation : Satellite RNAs are species of RNA associated with specific strains of some plant RNA viruses, although it is not necessary for their replication. Replication of this satellite RNA depends on the virus, so that it gets packaged with it to cause infection elsewhere. Therefore, satellite RNA depends on virus for its replication and transmission, even though it is unrelated to viral genome. Presence of sat-RNA leads to reduction in severity of disease symptoms, and therefore has been used for developing resistance against specific viruses.

Using the above principle, transgenic tobacco plants were produced which carried a DNA fragment, which when transcribed gives a species of satellite RNA which can reduce the severity of disease. Following two such attempts were made : (i) In one case, transgenic tobacco plants using DNA for RNA satellite of cucumber mosaic virus (CMV) were produced and when satellite free aggressive CMV strains were used for infection, the presence of satellite RNA (transcribed from DNA) in transgenic plants led to reduction in disease symptoms. These transgenic plants are also protected against another related virus i.e. tomato aspermy virus. In the other example involving tobacco ringspot virus (TobRV), transgenic tobacco plants could be produced, which carried DNA version of satellite RNA of tobacco ringspot virus (STobRV). These transgenic plants are protected against the deleterious effect of subsequent TobRV infection. The advantage of using satellite

RNA strategy over coat protein is that satellite RNA is not constitutively expressed like coat protein, but is expressed only after virus infection.

Antisense SAHH gene for resistance against a broad spectrum of viruses. In all the above approaches, a gene was used which conferred resistance against one specific virus.

However, if we could inhibit expression of a gene responsible for a non-essential host cell enzyme, required for virus replication, we can get resistance against a broad spectrum of viruses. In one such effort, the enzyme S-adenosyl homocysteine hydrolase (SAHH) was chosen for manipulation. SAHH is responsible for several methylation reactions including 5'-cap mRNA methylation, where S-adenosyl- methionine is used as methyl donor. If SAHH is inhibited, there will be under-methylation at the 5' end of viral mRNA leading to inhibition of viral replication. SAHH inhibitions have actually been used as antiviral agents and were found to suppress plant viruses in leaf disk assays. In a study reported in 1995, transgenic tobacco plants were produced which expressed antisense RNA against mRNA of a tobacco SAHH gene. This led to a reduction of SAHH level and consequent inhibition of viral replication/ infection in the transgenic plants.

Mutant Gene for a Protein Necessary for Cell to Cell Movement of Virus in the Host : In certain viruses (e.g. white colour mosaic virus or WC1MV) there are genes, which encode non-virion proteins that are necessary for cell to cell movement of the virus (but not necessary for viral replication). A mutation in one of these genes may hamper cell to cell movement of the virus. One such approach was used (reported in 1994), where a mutant version of a gene encoding WClMV-13kDa protein was used for producing transgenic plants. The transgenic plants were found to be resistant to several viruses. The mutant transgene was found to be dominant over the normal gene of virus infecting the transgenic host.

Resistance against Bacterial and Fungal Pathogens : Molecular basis for resistance against several diseases in a variety of crop plants has now been worked out. It has been shown in many cases that disease resistance is controlled by single resistance genes in the host that specifically recognizes pathogen strains containing complementary avirulence genes. A number of resistance genes from hosts and the corresponding avirulence genes from the pathogens have also been isolated and cloned in many cases. This has facilitated the development of disease resistant transgenic crop plants. Several examples are now available, where transgenic plants with resistance against bacterial and fungal pathogens have been produced. Some examples are given in Table.

Broad Spectrum Disease Resistance : The biochemical events responsible for disease resistance involve induction of a signaling cascade, which leads to hypersensitive reaction (HR) due to cell death in the host leading to inhibition of pathogen growth. The HR reaction depends on an interaction between *R* gene of the host and *Avr* gene of the pathogen, so that a broad-spectrum resistance through HR could be achieved due to combined expression of these genes (*R* gene and the *Avr* gene) in the same plant. However, expression of either one or both these genes needs to be tightly regulated to avoid harmful effect of their interaction on the host plant. Following two approaches were used:

(i) In one approach (two component sensor system), a pathogen-inducible promoter is attached with *Avr* gene, so that *Avr* was expressed only on attack of the pathogen, leading to HR;

(ii) In the other approach (four component genetically engineered acquired resistance or GEAR), a transposable element (TE) was inserted in *R* gene, so that on attack by the pathogen, the TE was excised and expression of R gene was restored leading to HR.

Some Pathogens for which Resistance has been Transferred in Some Crop plants.

Pathogen	*Disease*	*Resistance gene*	*Source of gene*	*Transgenic crop*
1. *Pseudomonas syringae* transferase gene	Wild fire	Acetyl	—	Tobacco
2. *Alternaria longipes* (soil bacterium)	Brownspot	Chitinase gene	*Serratia marcescens*	Tobacco
3. *Rhizoctonia solani* *Brassica napus*	Damping off	Chitinase gene	Bean	Tobacco and
4. *Phytophythora infestans*	Late blight	Osmotin gene	Potato	Potato
5. *Cercospora nicotianae*	Frogeye	Chitinase and glucanase genes	Rice & alfalfa	Tobacco

In summary, about two thirds of the genes, incorporated in the newly commercialized transgenic crops, confer resistance to biotic stresses including herbicide tolerance, insect resistance and virus resistance. The potential impact of these developments in terms of increased global food production and reduced insecticide use will be substantial, particularly with the insect resistant crops offering many new opportunities in crop protection.

Resistance Against Abiotic Stresses : Serious crop losses are also caused due to a variety of abiotic stresses, including drought, cold (frost), salinity, metal toxicity, etc. Transgenic plants resistant to these stresses have been produced, utilizing transgenes that alter the desired biosynthetic pathways, although even in the year 2002, these transgenic plants were only in a preliminary stage of development and perhaps will not be commercialized before 2005.

Resistance Against Stress (drought, salt, heat, freezing, etc.) : A number of genes are now known to be responsible for providing resistance against stresses such as heat, cold, salt and heavy metals. Stress tolerance in plants can be achieved in two ways: (i) improving protection from stress as done in oxidative stress, which is protected by super- oxide dismutases and other systems; (ii) reducing sensitivity to stress, as done in improving salt' tolerance by an osmolyte and chilling tolerance by increasing the level of polyunsaturated fatty acids. Studies are also being conducted on metabolites like proline and betaines that are implicated in stress tolerance.

The products of genes, whose expression is induced by stresses like drought, salt and freezing can be broadly classified into two groups: (i) Proteins that protect cells from dehydration; these include enzymes that are involved in the production of osmoprotectants, late embryogenesis-abundant (LEA) proteins, antifreeze proteins, chaperones and detoxification enzymes; (ii) Proteins that are involved in inducing transcription of stress responsive genes; these include transcription factors, protein kinases, and the enzymes involved in phosphoinositide metabolism. Genes encoding proteins belonging to both these groups have been utilized and will be utilized in future for the production of stress resistant transgenic plants. For instance, resistance against chilling was introduced into tobacco plants, by introducing a gene for 'glycerol-l-phosphate acyl transferase' enzyme from *Arabidopsis (Arabidopsis* is resistant to chilling). This enzyme, encoded by nuclear genome and later transported to chloroplast, determines the level of unsaturation of fatty acids in the phosphatidyl-glycerol of chloroplast membranes. The plants with high proportion of cis-unsaturated fatty acids (e.g. spinach, *Arabidopsis)* are resistant to chilling, and so are the transgenic tobacco plants carrying the gene for the above enzyme.

It has also been shown that a cis-acting promoter element, called dehydration response element (DRE) plays an important role in regulating gene expression in response to stresses like drought, salt and freezing. The transcription factors designated as DREB1A and DRE2B (DREB = DRE binding protein) specifically bind to DRE and activate transcription of genes containing the DRE sequence in *Arabidopsis*. The cDNAs for both these proteins have been isolated. Each of these cDNAs, along with CaMV35S promoter, was used for production of transgenic plants, which gave strong constitutive expression of stress inducible genes *(rd29A,~ kinl, cor6.6/kin2, cor47lrdl7, corl5a* and *erd10)* and were tolerant to stresses like drought, salt and freezing. Another cDNA encoding CRT/DRE binding protein (CBF1), when similarly used for the production of transgenic plants also led to increased freezing tolerance. Both in case of transgene *DREB1A* and *CBF1,* stress tolerance was associated with growth retardation under normal non-stress conditions. However, another stress inducible promoter *rd29A* was later used to drive *DREBJA,* so that the transgenic plants produced with this promoter exhibited stress tolerance and improved growth under non-stressed conditions.

Defense Against Water Stress : Accumulation of Osmoprotectants Like Glycine Betaine : Resistance against abiotic stresses like drought, salinity and cold (all leading to dehydration) is achieved in plants by the accumulation of osmotically active, low molecular weight non-toxic compounds, collectively described as osmoprotectants. These compounds help plants in two ways : (i) by acting as a cytoplasmic osmolyte, thereby facilitating water intake and water retention and (ii) by protecting and stabilizing macromolecules from damage induced by high salt levels. Some of these osmoprotectants include sugars, alcohols, proline and quaternary ammonium compounds (e.g. glycine betaine). Glycine betaine is a very effective osmolyte which accumulates during water stress due to high salinity, drought and low temperature in bacteria, cyanobacteria and in higher plants (particularly members of Chenopodiaceae and Poaceae, which are generally drought resistant). Several crops like potato, rice, tomato and tobacco do not accumulate glycine betaine, but can be made to accumulate it by manipulating glycine betaine biosynthetic pathway through transgenesis. Glycine betaine is obtained in two steps as follows:

$$\text{Choline} \xrightarrow{\text{CDH}} \text{betaine aldehyde} \xrightarrow{\text{BADH}} \text{glycinebetaine}$$

Biosynthesis of choline, which is a raw material for glycine betaine is under feedback control (not feedback inhibition), so that if choline is utilized by glycine betaine pathway introduced through genetic engineering, further choline uptake can be stimulated to allow synthesis of more glycine betaine.

The two enzymes involved in the above glycine betaine biosynthetic pathway include the following: (i) choline dehydrogenase (CDH) in *E. coli* and choline monoxygenase in spinach; each of which converts choline into betaine aldehyde; and (ii) betaine aldehyde dehydrogenase (BADH), which converts betaine aldehyde into glycine betaine. Bacterial CDH is the most useful enzyme, since it is able to catalyze not only the oxidation of choline to betaine aldehyde, but also the conversion of betaine aldehyde to glycine betaine. Tobacco lacks both the steps in glycine betaine biosynthetic pathway. Transgenic tobacco plants carrying *E. coli bet A* gene encoding CDH, were produced and were found to possess an 80% increase in salt tolerance (a report in 1996) at a salt (NaCl) concentration of 300 mM. Betaine aldehyde dehydrogenase (BADH) gene from *E. coli* was also used for production of transgenic plants exhibiting increased production of glycine betaine and imparting salt tolerance. Several other genes induced in response to salt stress in plants have been identified. These include *RAB, salt* and *osmotin* genes. The functions of these genes are not as clearly understood as those of the glycine betaine biosynthetic pathway. Transgenic plants carrying *E. coli* mannitol-1 -phosphate dehydrogenase gene have also been produced, but a rather high concentration of mannitol (100 mM) is needed to withstand high NaCl concentration. This concentration of mannitol can be deleterious to the plant. In view of the above, glycine betaine is considered to be one of the most efficient osmoprotectant. Low concentration of this osmolyte gives substantial protective effect in plants.

Defense Against Oxidative Stress using Superoxide Dismutase (SOD) Gene : An oxidative stress is frequently experienced by plant systems, because many stress factors induce the production of reactive oxygen causing oxidative damage within the cell. Protection against this damage is achieved through both enzymatic and non-enzymatic components. Key enzymes involved in this protection are superoxide dismutases (SODs), which convert superoxide redicals, into hydrogen peroxide. Due to their metal co-factors, three classes of SODs are recognized: (i) Cu/Zn SOD (generally found in the cytosol and chloroplasts and designated as Cyt Cu/Zn SOD and Chl Cu/Zn SOD respectively); (ii) Mn SOD (generally found in mitochondria) and (iii) Fe SOD (generally found in chloroplasts). Peroxidases and catalases operate closely with SODs for anti-oxidant defense mechanism, involving removal of H_2O_2 produced due to SODs. Inefficient removal of $H_2 O_2$ may lead to formation of hydroxyl radicals which are most reactive of all and therefore react with DNA, proteins and lipids. (H_2O_2 + O_2 + Fe^3~ + O_2 + OH). Overproduction of SOD has been shown to provide protection against (i) herbicides like 'paraquat', also called methyl viologen or Mv (a superoxide generating system); (ii) chilling stress and drought stress; and (iii) ozone damage as witnessed in tobacco and other members of Solanaceae. It was, however, noticed that enhanced SOD activity in the mitochondria had only a minor effect on ozone tolerance, while enhanced SOD activity in chloroplasts had 3 to 4 bold reduction in visible ozone injury. The most important work in this field has been done by M Van Montagu of Belgium. Another group working on SODs is led by E. Galun from Israel. Transgenic potato and tobacco have been produced utilizing Cyt and Chl Cu/Zn SOD and Mit Mn SOD. The genes for SODs were isolated either from tomato or from *Necotiana plumbaginifolia* and were integrated either in the nuclear genome or in the chloroplast /mitochondrial genomes.

Transgenic Plants with Chilling Resistance (desaturation of fatty acids) : Many of our crop plants (e.g. rice, maize, soybean) are injured or killed by exposure to low non-freezing temperatures in the range of 0-15°C. They are described as chilling-sensitive plants and carry lipids with only saturated fatty acids. It has been shown that presence of even one double bond in one of the 16 or 18 carbon saturated fatty acids found in the lipids, may lead to resistance against injury by low temperatures.

In higher plants, only three enzymatic activities are known, which are involved in desaturation of the saturated fatty acids: (i) *stearoyl-ACT desaturose,* located in the stroma of plastids, converts 18.O-ACP (acyl carrier protein) to 18:lc9-ACP (18.O means 18 carbon fatty acid with no double bond; similarly 18:1 means 18 carbon fatty acid with one double bond; c9 represents the position of double bond); (ii) the second membrane bound enzyme, located in plastids, desaturates 16:0 at *sn-2* in *phosphatidylglycerol (PG)* to 16:lc3; (iii) the third enzyme desaturates saturated 16:0 at *sn-2* in monogalactosyl diacylglycerol (MGDG) to 16:l07 (m-7-hexadecenoic acid). Except these three enzymatic activities leading to desaturation of fatty acids, no other fatty acid esterified to other lipids or to other positions of glycerol, can be desaturated in higher plants. Thus the scope of introducing double bond for resistance against low temperature becomes limited, if genes for higher plants only need to be exploited. Nevertheless some successful efforts have been made by utilizing genes from higher plants.

These efforts include the following: (i) saturation of the level of PG (a minor component of membrane lipids), was slightly reduced to 94%, in transgenic tobacco plants carrying a gene for *glycerol-3-phosphate acyltransferase* from *Arabidopsis thaliana,* which itself is chilling resistant. This, led to an increase in chilling resistance in transgenic plants in comparison to wild plants, (ii) Expression of plastidic w-3 desaturase from *A. thaliana* in the plastids of transgenic tobacco also led to a 10% increase of trienoic fatty acids over wild plant. This also imparted resistance against short-term exposure to chilling (1°C).

In the above examples of transgenic chilling resistant tobacco plants, genes from a chilling resistant higher plant *(A. thaliana)* were utilized. But the scope of using these desaturases from higher plants is limited as discussed above. Therefore, more recently (a report published in 1996), a broad-specificity A desaturase gene from a cyanobacterium *(Anacyctis nidulans)* was used, that was capable of introducing a cis-double bond at A9 position of both 16 and 18 carbon saturated fatty acids linked to membrane lipids. The gene was successfully introduced and its expression driven by CaMV 35S promoter. The enzyme was targeted into the plastids by the transit peptide of pea Rubisco small subunit. The transgenic tobacco plants thus produced had a highly reduced level of saturated fatty acid content in most membrane lipids and exhibited a significant increase in chilling resistance.

Improved Crop Productivity

Transgenic plants having improved crop productivity, in terms of harvest index or biomass production, have also been produced in model plant systems demonstrating the possibility of increasing plant productivity through transgenic approach. Two such examples will be briefly discussed, although the technology has not been commercially utilized so far.

Improved Harvest Index (overexpression of phytochrome A or PHY A gene) : The plant architecture in any crop is dependent on responses of the plant to signals from light. When plants in a population grow in close proximity to each other, the shade avoidance syndrome is invoked. In other words, many crop plants respond to crowding by growing taller. Although this response has competitive advantage in an ecological sense, this leads to reduced allocation of assimilates to leaf growth, storage organ production and reproductive development. Thus shade avoidance carries a penalty. Although it has adaptive significance, it leads to reduction in crop plant productivity due to non productive elongation of the stem.

In a crowded population of plants, photosynthetic pigments absorb most of the light in the visible region (i.e. 400-700 nm) but reflect or transmit radiation in the far red (FR) region (700-800 nm). FR reflection signals are perceived by the phytochromes {photoreceptors, whose two forms Pr and Pfr are interconvertible (Pr ⇌ pfr). The ratio of the two forms of phytochrome (Pr : Pfr) is a direct function of the ratio of red and far red light absorbed. In *Arabidopsis*, five phytochrome genes *(PHY A, PHY B, PHY C, PHY D, PHY E)* are known and the products of *PHY A* and *PHY E*, i.e. Phy A and Phy B exert mutually antagonistic effect on stem elongation as a shade avoidance response to far red. This suggested that artificial maintenance of Phy A at a high level in light grown plants might suppress shade avoidance response and thus may avoid the resource allocation for this wasteful growth from the point of view of crop productivity.

Transgenic tobacco plants when produced in 1996 for overexpression of Phy A, actually exhibited suppression of shade avoidance response to light that was artificially enriched in far-red light. For production of transgenic plants, the gene *PHY A*, cDNA, driven by CaMV-35S promoter was utilized. The product Phy A (which is labile to light and is quickly replaced by Phy B in wild plants) is stable in these transgenic plants. It can accumulate to high concentration, thus permitting high density crop, yet keeping dwarf. This led to increase in harvest index. This successful technology can now be used for other crop plants exhibiting shade avoidance response. In such crop plants (e.g., rice), the use of phytochrome gene for an increase of harvest index may lead to increased crop productivity.

Higher Biomass Production using Bacterial Haemoglobin Gene (VIIb) : Another significant research reported in early

1997 described successful production of transgenic plants that carry a gene encoding bacterial haemoglobin molecule *(VHb)*. These transgenic plants had higher growth rate, altered activity of metabolic pathways and increased productivity. The plants germinated earlier, grew faster and accumulated 80-100% more fresh weight after 35 days.

They also had more chlorophyll and nicotine. The gene was isolated from the obligate aerobic bacterium *Vitreoscilla,* (which produces haemoglobin under conditions of limiting O_2) and was fused to CaMV35S promoter. The haemiglobin protein produced in transgenic plants accounted for 0.1% of total leaf protein and was perhaps responsible for greater availability of O_2) as a substrate for a more productive metabolism requiring oxygen. The enhanced productivity due to haemiglobin reported in tobacco will certainly be tried in other crop plants in future.

Transgenic Plants for Hybrid Seed Production : During 1990-1992, an exciting example of producing transgenic plants with male sterility and fertility restoration genes became available in *Brassica napus.* This should facilitate production of hybrid seed without manual emasculation and controlled pollination as often practiced in maize.

In 1990, C. Mariani and others from Belgium, successfully used a gene construct having another specific promoter (from *TA29* gene of tobacco) and bacterial coding sequence for a ribonuclease *(barnase* gene from *Bacillus amyloliquefaciens)* for production of transgenic plants in *B. napus.* The results were spectacular in the sense that the transferred gene prevented normal pollen development leading to male sterility. The product of *barnase* gene is cytotoxic killing the tapetal cells, thus preventing pollen development. Utilizing this male sterility barnase gene construct (TA29— RNase), it was possible to introduce male sterility in other crops also. These crops include tobacco, lettuce, cauliflower, cotton, tomato, corn, etc.

Some Transgenic Plants Produced for Functional Food and Neutraceuticals.

Substance	*Benefit*	*Crop*	*Transgene*
Provitamin A	Anti-oxidant	Rice	Phytoene synthase, Phytoene desaturase, Lycopene cyclase
Vitamin E	Anti-oxidant	Canola	y-tocopherol methyl transferase
Flavonoids	Anti-oxidant	Tomato	Chalcone isomerase
Fructans	Low calorie	Sugarbeet	l-sucrose:sucrose fructosyl transferase
Iron	Iron fortification	Rice	Ferritin, metallothionein, phytase

The same group of workers *(Mariani et al, 1992)* used another gene construct later involving the same anther specific promoter i.e. *TA29* and the *barstar* gene from *B. amyloliquefaciens,* for production of transgenic plants in *B. napus.* The product of *barstar* gene is a ribonuclease-inhibitor. It forms a complex with ribonuclease and neutralizes its cytotoxic properties. In *Bacillus,* the ribonuclease is active extracellularly and the bacterium itself is protected by a ribonuclease inhibitor protein coded by *barstar* gene. When transgenic male sterile plants (with *barnase* gene) were crossed with transgenic male fertile plants (with *barstar* gene), the F1 plants expressed both genes so that male fertility was restored due to suppression of cytotoxic ribonuclease activity in the anther by the formation of cell specific RNase/RNase-inhibitor complexes. This system of transgenic plants should facilitate hybrid seed production in crop plants in general. In India, barnase and baistar genes have been successfully transferred to *Brassica guncea* at University of Delhi South Campus (UDSC). These may eventually be used for hybrid seed production.

Improved Nutritional Quality (Functional Food and Neutraceuticals) : Transgenic crops have also been produced to provide for functional food and neutraceuticals. Functional food is defined as the food consumed as a part of normal diet, but would have, in addition to basic nutritional function, physiological benefit and/or reduced risk of chronic diseases. Neutraceuticals, on the other hand, are products isolated or purified from foods that are generally sold in medicinal forms (biopharmaceuticals). Examples of transgenic plants developed for functional food and neutraceuticals include those that would have reduced concentration of life threatening substances like several allergens, and will also have high concentration of other health promoting nutrients including improved quality of storage proteins (e.g., food grain, rich in lysine and sulphur rich amino acids), altered fatty acid composition in edible oil, a variety of vitamins, anti-oxidants, cholesterol lowering margarine with plant sterols, low calorie polysaccharides and iron fortified food (Table).

Transgenic Plants for Oil Production (rapeseed and other oil crops) : The four major globally important oil crops are soybean, oil palm, rapeseed and sunflower, which account for -72% of world's vegetable oil production. Considerable progress in engineering oil production particularly in rapeseed, has been made in recent years (1990s). Although in 1980s, major emphasis was laid on the development of transgenic crops for herbicide resistance, virus resistance and insect resistance, the emphasis in 1990s shifted towards resistance against abiotic stresses and towards improvement of quality traits like oil and protein content and their quality. Consequently in the year 1993, crops containing modified quality traits (seed oil or protein composition) accounted for 31% of all transgenic rapeseed releases (field trials). The improvement in oil content and oil quality through gene-transfer technology in rapeseed was easier relative to other quality traits in some crops, due to the following reasons: (i) simple transformation system; (ii) involvement of one or few genes; (iii) relatedness with *Arabidopsis* whose genome has been thoroughly investigated leading to isolation of many genes including those for oil production. Similar genes have also been isolated from other oil producing crops (e.g. castorbean). Some of these *Arabidopsis* genes have already been utilized for rapeseed improvement.

Oil Content in Seeds : Major oilseed crops contain about 20-45% (w/w) oil in their seed. Further increase has been difficult, although there are crops having 60-70% of oil in their seed (e.g. candlenut, sesame, oiticia, ucuhuba, etc). Rapeseed also contains twice as much oil as soybean, and this is attributed to higher rate of oil synthesis in rapeseed than in soybean. It is therefore assumed that if flux of carbon towards oil is upregulated by manipulating one or two enzymes, more seed oil may accumulate.

Proportion of Desired Polyunsaturated Fatty acids (PUFAs) : A desired fatty acid may be made to constitute 30-40% of the seed-oil, but further increase in the level of such a novel fatty acid has been difficult. Single gene insertions have actually been shown to alter the fatty acid (e.g., lauric acid) composition in rapeseed. Transfer of another gene encoding *sn-2* acyltransferase led to increase of this fatty acid to 60%. However, even this high level of desired fatty acid may be inadequate. For instance, there is a demand in rapeseed for increasing the erucic acid content from 50% to 90% of total seed oil, which is difficult to achieve (although for edible oil, low or zero erucic acid is desired, but for industrial purposes, high level of erucic acid is necessary). However, in castorbean, varieties are available where a single fatty acid constitutes up to 90% of seed oil, suggesting that it is not impossible to achieve this level in rapeseed and other seed oil crops (see later for details).

By-products from Oil-seed Crops : Several by-products are produced either during oil extraction or during the industrial uses of oil-seeds: (i) Protein-rich meal is produced during oil extraction from oil seed with anti-nutritional agents like glucosinolates. The demand for this meal for animal feed has increased. There is a hope that glucosinolate metabolism can be manipulated to improve the nutritional value of this meal. Such an improvement may be essential, because it may not be economically viable to grow rapeseed for industrial use alone, unless protein-rich meal obtained as a by-product can find a ready market, (ii) Glycerine is produced as a by-product during the industrial uses of oil seed for the manufacture of detergents, polymers and biodiesel. The availability of glycerine in increased quantities in world market led to the development of its new uses (e.g., soft drinks, fabric conditioners, de-icing and drilling fluids). This has also led to an increased demand for glycerine, (iii) *Oleosins* are proteins that surround oil-storage bodies in seeds as monomolecular layers. These proteins have been re-engineered to act as carriers of pharmaceutical peptides such as *thrombin* or *hirudin.* which are thus targeted to oil bodies and can be easily separated from oil bodies. Several industrial enzymes (e.g., cellulases, proteases, Upases, etc) can also be targeted to oil bodies in a similar manner. This technology is being developed in Canada and will allow simultaneous recovery of the novel peptides/proteins of interest along with seed oil and protein-rich meal.

Improvement of other Oil-seed Crops : The mechanism of fatty acid biosynthesis and modification is common to all oil producing crops. However, in oil-rich fruit crops such as palm, coconut and avocado, oil is stored in a different way since it is used as an attractant for animals (fruit-oils accumulate as large irregular droplets of 20 ^m, while seed oils are deposited as compact membrane-bound spheroidal droplets of 1-2 Jim). Due to this common mechanism, same results were obtained when same lipid-related transgenes were transferred to soybean or rapeseed. Similar results would also be expected in palm, coconut, etc. Therefore, available genes for improvement of oil can be used in other oil-seed crops and oil-rich fruit crops such as linseed and palm oil. Transformation technology for rapeseed and soybean have already been developed and those for sunflower and linseed were still being developed in the year

1995. Therefore, eventually it should be possible to transform all the major annual oil-seed crops. The transformation of the perennial oil crops, such as oil palm and olive is much more challenging, but the demand of oil palm will compel the development of transgenic palms for increased oil content and improved oil quality, particularly in major palm oil producing country, Malaysia. More genes involved in oil production will be isolated and utilized in a variety of oil producing crops in the coming decades.

Altered Fatty Acid Composition in Brassica and Soybean seed Oil : Oils and fats are tryglyceride esters of glycerol involving aliphatic straight chain fatty acids, which may be either (i) saturated, with no double bonds, or (ii) unsaturated, with 1, 2 or more double bonds. The properties of oil or fat depends on the number of carbon atoms and the number of double bonds that are present in the constituent fatty acid(s). The melting point of oil or fat decreases with the decrease in the number of carbon atoms and increase in the number of double bonds, so that the oils and fats with fewer carbon atoms and more double bonds will have relatively low melting point, which in its turn determines the oil quality.

Therefore, fatty acids are characterized by the number of carbon atoms and the double bonds, which are generally given in parentheses after the name of a fatty acid. For instance, saturated fatty acids like stearic acid (18:0), palmitic acid (16:0) and lauric acid (12:0) respectively have 18, 16 and 12 carbon atoms with no double bonds. Similarly, the unsaturated fatty acids like oleic acid (18:1), linoleic acid (18:2) and linolenic acid (18:3), each has 18 carbon atoms, but differ in having 1, 2 and 3 double bonds respectively. Erucic acid (22:1) that is common in *Brassica* seeds and is often responsible for heart diseases, has 22 carbon atoms and a single double bond. While oils and fats, with high proportion of saturated fatty acids, and derived from palm, coconut and cacao, are important for industrial uses, those with high proportion of unsaturated fatty acids (derived from oilseed crops (soybean, sunflower, groundnut, sunflower, sesame, cotton, etc.) are used as cooking oil, margarine, salad oil, etc. In edible oils, relative proportion of saturated fatty acids and unsaturated fatty acids can be altered using transgenic approach. In the past, specialized fatty acid composition in the seed oil, desired for edible and industrial purposes, has been achieved by conventional plant breeding or by mutagenesis programmes. Following are some examples: (i) removal of erucic acid from rapeseed oil to create 'canola'; (ii) reduction in linolenic acid content in flax; (iii) increase in stearate content (six times that in wild type) in sunflower (upto 12% stearate) and soybean (upto 30% stearate). Tissue specific antisense RNA expression has recently been used for reducing 'stear'oyl-ACP desaturase' (ACP = acyl carrier protein) activity in seeds, leading to alteration in the ratio of saturated to unsaturated fatty acids. Due to tissue specific expression of anti-sense RNA, integrity of membrane lipids in leaf remained unaffected. 'Stearoyl-ACP desaturase' catalyzes the first desaturation step in seed oil biosynthesis, converting 'stearoyl-ACP' to 'oleoyl- ACP.' Seed specific anti-sense gene constructs of *B. rapa's (syn-B. campestris)* stearoyl-ACP desaturase were used for production of transgenic *B. rapa* and *B. napus* plants.

This led to reduction in desaturase activity, resulting into a dramatic increase in the level of stearate in the seeds. High stearate *B. napus,* though unsuitable as a cooking oil, was developed to provide for an alternative source of saturated fatty acids, which are generally obtained from cacao butter. This high level of saturated fatty acids, due to high melting point of their corresponding fats, are useful for getting semisolid products like margarine and shortenings (fats for pastries, etc.).

High laurate transgenic canola (rich in lauric acid, which is a saturated fatty acid with 12:0 constitution) has also been produced and commercialized as a variety named 'Laurical'. Oil processed from this transgenic canola is similar to coconut and palm oils, and is being sold to industry for use in chocolate candy coatings, coffee whitener, icings, frostings, and whipped toppings. This oil is also used in cosmetic industry.

In contrast to the above example, where the proportion of saturated fatty acids (stearate, laurate) was increased through genetic engineering, there are other examples, where the proportion of unsaturated fatty acids was increased, because according to health experts, for cooking oil, monounsaturated fats should be preferred over saturated fats found in beef, pork, cheese and other dairy products.

In soybean, while the oil processed from the conventional soybeans have an oleic acid (a monounsaturated fatty acid) content of 24%, that processed from the new transgenic soybeans has been shown to have oleic acid content, as high as 80%. Oil produced from these varieties is similar to that of peanut and olive oils. High oleic acid was obtained through anti-sense technology by blocking the activity of olyl desaturase enzyme, which converts oleic acid into linoleic acid.

Lysine Rich Transgenic Crops (canola and soybean) : Lysine rich transgenic canola *(Brassica)* and soybean *(Glycine max)* were produced in 1995 by researchers at DuPont and University of Delaware (USA) using the technique of metabolic engineering.

Crops, Rich in Vitamin A ('Golden Rice' and 'Golden Mustard') : Worldwide, it is estimated that 124 million children suffer with vitamin A deficiency (VAD), which annually causes irreversible blindness in 500,000 children and results in large scale deaths of these children. In Southeast Asia, due to VAD, every year five million children develop symptoms that may lead to blindness. VAD is also a matter of concern in several areas of Africa, Latin America and the Caribbean. It is estimated that vitamin A nutrition may prevent approximately 1 to 2 million deaths every year among children aged 1 to 4 and an additional 0.25 to 0.5 million during higher age group of children. In view of these concerns, transgenic crops like rice and mustard, described as 'golden rice' and 'golden mustard' that are rich in vitamin A have either been developed or are being developed (the term 'golden' refers to the golden colour of the rice grain or mustard oil due to vitamin A). Golden rice was developed by using two genes from *Narcissus* (phytoene synthase = *psy;* lycopene P-cyclase = *Icy)* and one gene from *Erwinia* (phytoene desaturase = *Crtl).*

Extensive research during 1996-2000, by Ingo Potrykus in Switzerland and Peter Beyer in Germany, led to the development of transgenic 'Golden Rice', containing beta-carotene and carotenoids that are precursors of Vitamin A (Fig.). It is believed to be one potential solution to VAD, although improved diet including green vegetables, animal products and vitamin pills could be other possible alternative solutions. In the year 2000, efforts were made to donate golden rice to alleviate poverty and poor health in the third world. In this connection several private companies having proprietary rights (intellectual property rights) on technologies used in this research licensed these technologies free of charge for humanitarian uses of golden rice in developing countries. During the year 2001, sample seed of 'Golden Rice' were also provided to International Rice Research Institute (IRRI) in Philippines, to investigate its safety and utility. It should be ready for field trials by 2003-04 and for the farmers, fields by 2005-06.

In December 2000, a collaborative project to develop 'Golden Mustard' was launched between the Tata Energy Research Institute (TERI) in India, and Monsanto and Michigan State University (MSU) in USA. 'Golden Mustard' will yield cooking oil with high content of beta-carotene (pro-vitamin A). The technology thus will provide rice and mustard with high level of vitamin A and in future, the technology will certainly be extended to other crops like maize.

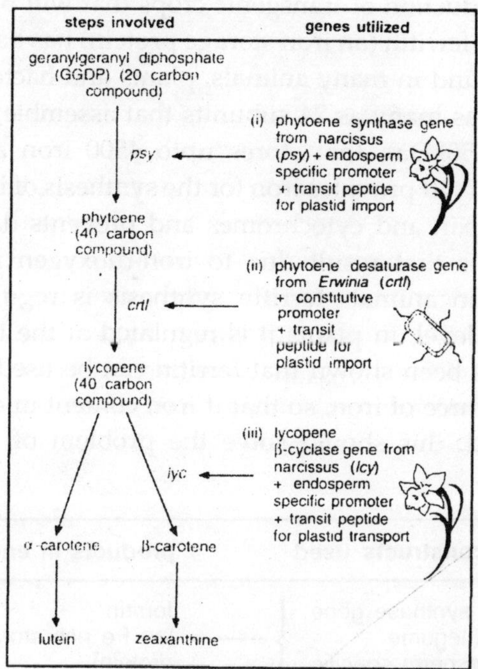

Steps involved, sources of genes and important components of gene constructs, used for the development of pro-vitamin A rice.

Transgenic Crops Fortified with Iron

Iron deficiency is also a serious nutritional problem, affecting an estimated 30% of the world population. Programs involving supplementary food and tablets to overcome iron deficiency are difficult to implement in developing countries, due to the costs involved and due to limited number of health care programs available there. Although there are some crops like spinach and some leguminous crops that have a relatively higher iron content, they contain oxalic acid and phytate-like substances, which reduce the bioavailability of iron. In view of this, efforts have been made to develop rice crop, which will allow fortification of iron in the grain with enhanced level of its bioavailability.

For production of transgenic crops that will produce food rich in iron, ferritin (an iron storage protein) has been targeted, which is found in many animals, plants and bacteria. Ferritin consists of as many as 24 subunits that assemble into a large complex (450kDa), that stores upto 4500 iron atoms in its central cavity. It provides iron for the synthesis of iron proteins like ferredoxin and cytochromes and prevents damage from free radicals that result due to iron-dioxygen interactions. Although, in animals ferritin synthesis is regulated at the translation level, in plants it is regulated at the transcription level. It has been shown that ferritin can be used by animals as a rich source of iron, so that if iron content in cereal grains is increased, this should solve the problem of dietary iron deficiency.

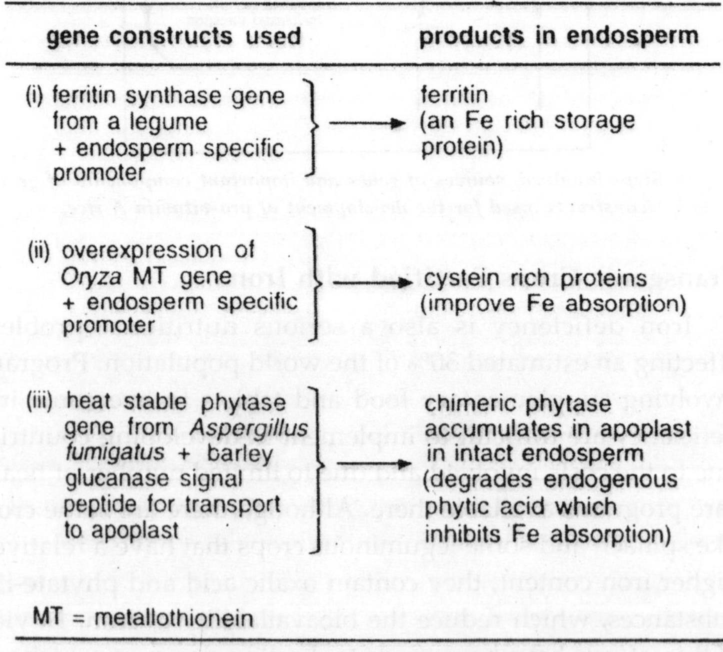

gene constructs used		products in endosperm
(i) ferritin synthase gene from a legume + endosperm specific promoter	→	ferritin (an Fe rich storage protein)
(ii) overexpression of *Oryza* MT gene + endosperm specific promoter	→	cystein rich proteins (improve Fe absorption)
(iii) heat stable phytase gene from *Aspergillus fumigatus* + barley glucanase signal peptide for transport to apoplast	→	chimeric phytase accumulates in apoplast in intact endosperm (degrades endogenous phytic acid which inhibits Fe absorption)

MT = metallothionein

Strategies used for development of high-iron rice.

For production of rice crop, having seed fortified with iron, genes for ferritin protein, isolated from soybean and *Phaseolus*, have been transferred to rice in two independent studies, one conducted in Japan and reported in the year 1999 and the other conducted in Europe, and reported in the year 2001.

A rice variety (IR68144) was developed at IRRI by crossing two varieties, and was rich in Fe and Zn. In 1999, 27 religious sisters used this rice for months and showed an improved iron status. Later, in one of the largest feeding trials of a staple food, 300 nuns from 8 convent schools in the Philippines capital Manila helped in testing a rice anaemia, which affected one-third of the world population. In Asia it affected 60% of pregnant women and 40% of school children. About 150 nuns used IR68144, and other half used the normal rice; a significant difference was observed.

Transgenic Plants suitable for Transport and Long Shelf-life : A number of examples are available, where transgenic plants suitable for food processing have been developed, (i) Bruise resistant tomatoes were developed which express antisense RNA against polygalacturonase (PG) that attacks pectin in the cell walls of ripening fruit and thus softens the skin (Fig.). (ii) Tomatoes and cantaloupe melon fruits exhibiting delayed ripening were developed, either by using antisense RNA against enzymes involved in ethylene production (e.g. ACC synthase or ACC oxidase), or using gene for ACC deaminase, which degrades 1 amino-cyclopropane-1 carboxylic acid (ACC), an immediate precursor to ethylene. This will increase the shelf life of tomato. The tomato and cantaloupe melons expressing antisense ACC synthase or ACC oxidase gene will exhibit delayed ripening of fruits that are still attached to the plant. However, in tomatoes expressing ACC deamainase gene, fruits displayed delayed ripening when detached from the plant and allowed to ripen on the shelf. When fruits are allowed to stay on the plant longer, they accumulate higher amounts of soluble sugars, thus achieving better flavour.

Tomatoes obtained thus (using ACC synthase) were named 'Flavr Savr'. (iii) Tomatoes with elevated sucrose and reduced starch could also be produced using sucrose phosphate synthase gene, (iv) Tomatoes with high level of total solids and soluble solids were produced using isopentenyl transferase *(ipt)* gene from *Agrobacterium tumefaciens*. This is a key enzyme in the biosynthesis of plant growth regulator, cytokinin. Expression of the gene was modulated using a promoter permitting ovary specific expression. This led to increase in cytokinin level in tomato ovaries giving higher level of solids, (v) Starch content in potatoes could be increased by 20-40% by using a bacterial ADP glucose pyrophosphorylase gene *(ADP GPPase)*.

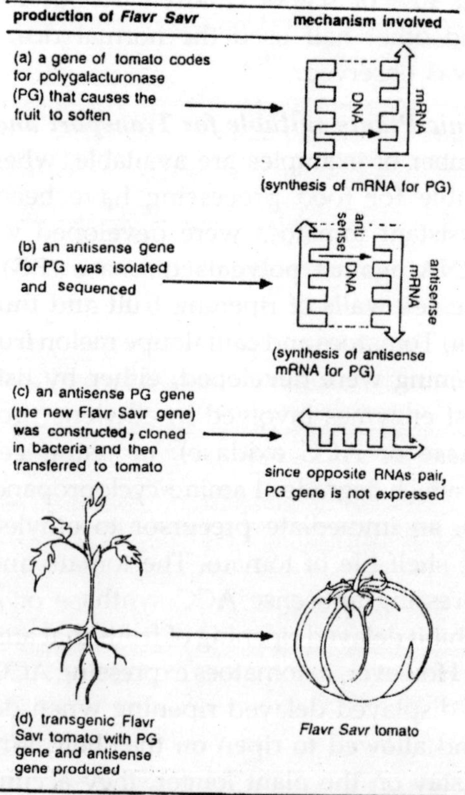

Production of transgenic 'Flavr Savr' tomato using an antisense RNA gene.

Transgenic Plants for Floriculture

During 1990s, the interest in the production of transgenic ornamental plants also gained momentum. While in 1990, transformation procedure was available for only one ornamental i.e. *Petunia,* in 1995-2000, transformation procedures became available for many ornamentals including some monocots. These cut-flower species include rose, carnation, tulip, chrysanthemum, gerbera, daffodils, gypsophilla, lily, iris, etc. In several of these cut flowers, recent work led to successful production of transgenics with novel aesthetic properties including novel colours, fragrance, plant and flower architecture, longer life, etc. Some of these traits having a commercial demand have been the target for genetic engineering and will be briefly discussed.

Plant Size : The ratio of the concentrations of cytokinin and auxin can be altered in transgenic plants through expression of oncogenes derived from Ti/Ri plasmids. Following are some examples :

(i) Constitutive expression of *role* gene encoding cytokinin-(3-glucosidase from Ri plasmid leads to release of 'active cytokinins giving desired dwarf and bushy phenotype (e.g. potato becoming an ornamental);

(ii) GA concentration can be altered, so that reduced GA can give dwarf plants and increased GA can give tall plants; two genes *(gal, anl)* involved in GA biosynthesis can be used to alter the levels of GA; overexpression of phytoene synthase gene also leads to reduced GA level and dwarfism.

Flower Colour : The first application of genetic engineering to modify flower colour led to the production of an orange pelargonidin-producing *Petunia* variety, which produced flowers of pale brick red colour. This was achieved by the expression of the dihydroflavonal-4 reductase gene *(dfr)* from maize in a petunia line. The variety was unstable due to transgene silencing and was therefore not viable in commercial market. Subsequently, several transgenic plants with *dfr* gene were used for breeding with elite material to produce a commercially viable brown orange petunia variety in the Netherlands.

Production of blue flower in rose, tulip or carnation has also been an excitement among biotechnologists and floriculturists. It is complicated, because it needs following three factors : (i) synthesis of 3', 5' hydroxylated anthocyanins (delphinidins); (ii) presence of flavonal co-pigments and (iii) relatively high vacuolar pH. Delphinidins are not available in roses, but two genes responsible for blue colour that have been isolated from petunias include the following: (i) gene for enzyme flavonoid-3'-5' hydroxylase (for synthesis of delphinidins) and (ii) gene for flavonal synthase (for synthesis of flavonals). The genes that control vacuolar pH were difficult to clone but some progress has been made. From six genes *(pHl-pH6), pH6* had been cloned, and *pH3* and *pH4* were being cloned in 1995. Three other genes involved in regulation of anthocyanin biosynthetic pathway i.e. *anl, an2* and *anil* also cause alterations in intracellular pH and can be isolated by differential screening of cDNA library.

Chalcone synthase *(Chs)* is another gene which has been used for production of pink, white and variegated flowers (sense and antisense genes were used) in petunias, chrysanthemum, gerbera and roses. Transcriptional activators of anthocyanin biosynthetic pathway, namely R and Cl have also been used to alter the intensity of flower colour. Mutations in regulatory sequence also gave novel pigmentation patterns in ornamentals like snapdragon. The gene *Chs* has also been used to produce white moneymaker chrysanthemum.

Architecture of Flower and Inflorescence : The identity of floral organs can be altered by using homeotic genes involved in determining the identity of floral whorls and the number of organs in each whorl. Mutants in *Petunia* and *Arabidopsis* have been isolated, which have altered number of floral organs. These can be used to change the architecture of the flower.

The inflorescence in ornamentals may have determinate (terminal flowers) or indeterminate growth (indefinite number of flowers). A single gene is sufficient to convert the inflorescence from indeterminate to determinate type or vice versa. Several of these genes have already been cloned.

Flowers with Longer Life

The vase life of flowers can be altered by manipulating the biosynthesis of ethylene (Fig.). The enzymes ACC synthase and ACC oxidase are encoded by genes *acs* and *aco* respectively and both have been cloned from many species including carnation. Transgenic carnation having antisense *aco* gene or sense *acs* gene have been produced and exhibited longer vase-life (delay in petal senescence). They have already been approved for commercial release. Other strategies for longer vase-life of flower include the following : (i) overexpression of a gene for ACC deaminase, that metabolises ACC before it is converted to ethylene (Fig.); (ii) overexpression of the gene for SAM hydrolase (isolated from T3 phage); (iii) expression of a gene for isopentenyl transferase *(ipt)* giving increased level of cytokinin; (iv) isolation and use of genes involved in ethylene signal transduction pathway.

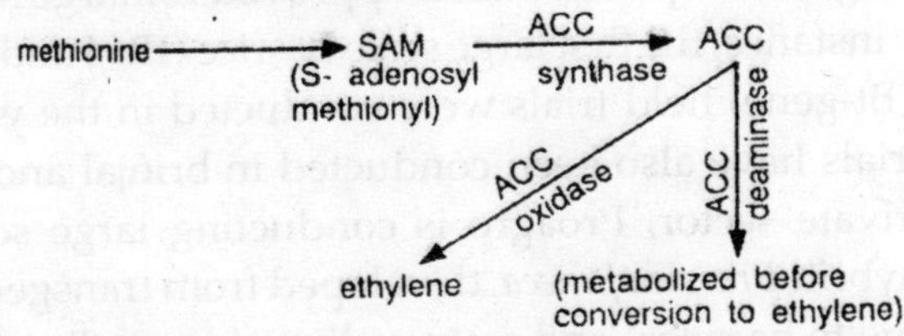

Biosynthesis of ethylene and its manipulation for longer vase life.

Flower Fragrance : Monoterpens are an important class of compounds that are involved in fragrance and are commercially valuable as essential oils for perfumes. These are derived from geranyl pyrophosphate by various monoterpene synthases, and a gene for one of them (S-linalool synthase) was cloned during early 1990s. Two other genes modifying monoterpenes were also cloned. These genes will be used in genetic engineering to modify floral fragrance.

Transgenic Crops in India : Several transgenic crops have been developed in India or imported from abroad either for direct field testing or for involving them in plant breeding programmes to develop varieties that are adapted to Indian conditions, but carried transgenes.

Transgenic Crops being Developed : In several laboratories in India, transgenic crops have been successfully produced. In still other cases, transgenic work is still in progress and transgenic plants will soon become available. The details of the laboratories where this work is in progress, at the beginning of the year 2002, are given in Table. Programmes are also under way to develop transgenic pulse crops including chickpea and pigeonpea.

Transgenic Crops Field-tested : Among crops that were field tested in India during 1996-2001, the main crops are cotton, brinjal, tomato and rice. While Bt-cotton, that was grown without approval in large areas (> 10,000 hectares) in Gujarat in the year 2001, was later field tested (Mahyco-Monsanto) and approved for commercial cultivation in the year 2002. Several other transgenic crops have been approved for large scale field trials. For instance, the first large scale Bt -rice (IR64 and Basmati, carrying Bt-gene) field trials were conducted in the year 2001. Similar trials have also been conducted in brinjal and tomato. In the private sector, Proagro is conducting large scale field trials of hybrid *Brassica juncea*, developed from transgenic male-sterile (due to barnase) and restorer lines (due to barstar). This hybrid mustard could not receive approval for commercial cultivation in the year 2002, but should receive this approval in 2003-04.

Transgenic lines in advanced stage of development or field testing in India.

Institute	Crop	Transgene inserted	Aim
Central Tobacco Research Institute	Tobacco	*Bt* toxin genes	Resistance to caterpillar (*Spodoptera litura*)
CPRI Shimla	Potato	*Bt* toxin gene	Resistance to tuber moth (*Phthorimaea opurculella*)
IARI, New Delhi	Brinjal, rice, tamato	*Bt* toxin gene	Resistance to insects
	Tomato	ACC synthase gene	Delayed fruit ripening
	Brassica juncea	Anniexin gene	Tolerance to moisture stress
	Potato	Osmotin	Tolerance to moisture stress
DRR, Hyderabad	Rice	*Bt* toxin gene	Resistance to yellow stem borer
	Rice	Chitinase gene	Resistance to sheath blight disease
Bose Institute, Calcutta	Rice	*Bt* toxin gene	Resistance to yellow stem borer
Delhi University South Campus,	Mustard/rapeseed	Barnase and barstar	Hybrid
JNU, New Delhi	Potato	*Ama-1* gene from *Amaranthus*	To improve nutritional quality
Central Institute of Cotton Research, Nagpur	Cotton	*Bt* toxin gene	Resistance to insect pests
M/s Proagro PGS India Ltd., New Delhi	Mustard rapseed, brinjal, cauliflower, cabbage, cauliflower	Barnase and barstar	Hybrid development
	Tomato	*Bt* toxin gene	Resistance to insect pests
M/s Mahyco, Mumbai	Cotton	*Bt* toxin gene	Resistance to insect pests

Data mainly based on information presented by Ghosh and Ramanaiah (2000)

7

Transfer Methods

For production of transgenic animals, DNA is usually microinjected into pronuclei of embryonic cells at a very early stage after fertilization, or alternatively gene targeting of embryo stem (ES) cells is employed. This is possible in animals due to the availability of specialized *in vitro* fertilization technology, which allows manipulation of ovule, zygote or early embryo. Such techniques are not available in plants. In contrast to this, in higher plants, cells or protoplasts can be cultured and used for regeneration of whole plants. Therefore, these protoplasts can be used for gene transfer followed by regeneration leading to the production of transgenic plants. Besides cultured cells and protoplasts, other meristem cells (immature embryos or organs), pollen or zygotes can also be used for gene transfer in plants. The enormous diversity of a plant species for its response to gene transfer techniques and the availability of diverse genotypes in each species, made it necessary to develop a variety of techniques, suiting different situations. These different methods of gene transfer in plants will be discussed in this chapter.

Target Cells for Transformation

The first step in gene transfer technology is to select cells that are capable of giving rise to whole transformed plants. Transformation without regeneration and regeneration without transformation are of limited value. In many species, identification of these cell types is difficult. Some of these cell types and the methods used for regeneration of whole plants from them are listed in Table.

As can be noticed in Table, cultured cells or protoplasts are often used for gene transfer. This is unlike the situation in animals, because the plant cells are totipotent and can be stimulated to regenerate into whole plants *in vitro* via organogenesis or embryogenesis. However, *in vitro* plant regeneration imposes a degree of 'genome stress', especially if plants are regenerated via a callus phase. This may lead to chromosomal or genetic abnormalities in regenerated plants a phenomenon referred to as somaclonal or gametoclonal variation. In contrast to this, gene transfer into pollen (or possibly egg cells) may give rise to genetically transformed gametes, which if used for fertilization (in vivo) may give rise to transformed whole plants. Similarly, insertion of DNA into zygote *(in vivo* or *in vitro)* followed by embryo rescue, may also be used to produce transgenic plants. Another alternative approach is the use of individual cells in embryos or meristems, which may be grown *in vitro* for the production of transgenic plants.

Theoretical Routes for Production of Transgenic Plants.

Target cell types	Method used for regeneration
1. Cultured cells or protoplasts	Organogenesis or embryogenesis via callus phase
2. Meristem cells from immature embryo or organ	*In vitro* plant regeneration from transformed cells

Contd....

Target cell types	*Method used for regeneration*
3. Cells in immature embryos, shoot and flower meristems	Normal development *(in vivo)* of embryo, shoot or flower followed by use of pollen (transformed) from chimaeric plant to produce transformed seed
4. Pollen	Pollen (treated with DNA) used for pollination leading to the production of transgenic plants
5. Zygote	Direct development *(in vivo)* of transgenic plants

Vectors of Gene Transfer

General Features : The common features of vectors used for transformation of eukaryotic cells are shown in Figure. Most vectors carry marker genes, which allow recognition of transformed cells (other cells die due to the action of an antibiotic or herbicide) and are described as selectable markers. A number of selectable markers are now known and are listed in Table. Among these marker genes, the most common selectable marker gene is *nptII,* providing kanamycin resistance.

Other common features of a suitable transformation vector include the following:

(i) multiple unique restriction sites (a synthetic polylinker);

(ii) bacterial origins of replication (e.g. *ColE1).*

The vectors having these properties may not necessarily have features, which facilitate their transfer to plant cells or integration into the plant nuclear genome. Therefore, *Agrobacterium* Ti plasmid is preferred over all other vectors, because of wide host range of this bacterial system and the capacity to transfer genes due to the presence of T-DNA border sequences (see below).

Vectors Based on Ti and Ri Plasmids of *Agrobacterium.*

Structure and Function of Ti and Ri Plasmids : The most commonly used vectors for gene transfer in higher plants are based on tumour inducing mechanism of the soil bacterium *Agrobacterium tumefaciens,* which is the causal organism for crown gall disease. A closely related species, *A. rhizogenes* causes hairy root disease. An understanding of the molecular basis of these diseases led to the utilization of these bacteria for developing gene transfer systems. It has been shown that the disease is caused due to the transfer of a DNA segment from the bacterium to the plant nuclear genome.

The DNA segment, which is transferred is called T-DNA and is part of a large Ti (tumour inducing) plasmid found in virulent strains of *Agrobacterium tumefaciens.* Similarly Ri (root inducing) megaplasmids are found in the virulent strains of *A. rhizogenes.* The Ti and Ri plasmids, inducing 'crown gall disease' and 'hairy root disease' respectively have been studied in great detail. However, we will discuss only those aspects of these plasmids which are relevant to the design of vectors for gene transfer in higher plants.

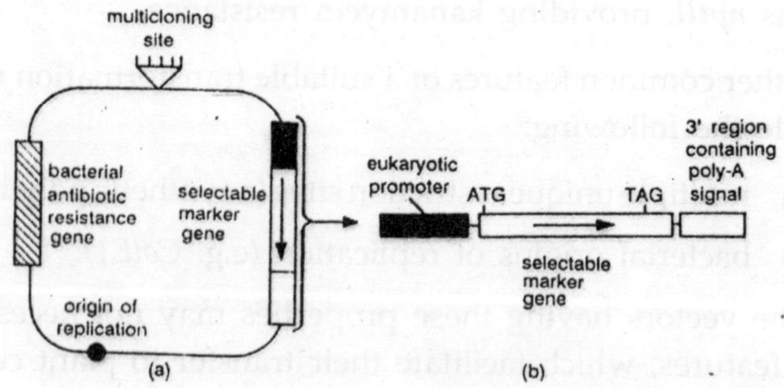

(a) Generalised vector for the transformation of eukaryotic cells; (b) details of the selectable marker genes (Redrawn from D. Grierson, 'Plant Gentic Engineering', 1991).

Selectable Marker Genes in Vectors Used for Gene Transfer.

Selectable marker gene	Substrate used for selection
I. Antibiotics.	
1. Neomycin phosphotransferase *(npt II)*	G418, kanamycin, neomycin
2. Hygromycin phosphotransferase *(hpt)*	Hygromycin B
3. Dihydroforate reductase *(dhfr)*	Methotrexate trimethoprim
4. Gene *ble* (enzyme not known)	Bleomycin
5. Gentamycin acetyltransferase	Gentamycin
6. Streptomycin phosphotransferase	Streptomycin
II. Herbicides	
7. Mutant form of acetolactate synthase *(als)*	Chlorosulfuron imidazolinones
8. Bromoxynil nitrilase	Bromoxynil
9. Phosphinothricin acetyltransferase *(bar)*	L -phosphinothricin (PPT; also known as bialaphos)
10. 5- enolpyruvylshikimate -3 -phosphate (EPSP) synthase *(aro·A)*	Glyphosate (Roundup)
III. Others	
11. Phosphomannose isomerase *(pmi)*	Mannose

Most Ti plasmids have four regions in common, which are shown in Figure.

(i) Region A, comprising T-DNA is responsible for tumour induction, so that a mutation in this region leads to the production of tumours with altered morphology *(shooty* or *rooty* mutant galls). Sequences homologous to this region are always transferred to plant nuclear genome, so that the region is described as T-DNA (transferred DNA).

(ii) Region B is responsible for replication.

(iii) Region C is responsible for conjugation.

(iv) Region D is responsible for virulence, so that mutation in this region abolishes virulence.

This region is therefore called virulence *(vir)* region and plays a crucial role in the transfer of T-DNA into the plant nuclear genome. The components of this Ti plasmid have been used for developing efficient plant transformation vectors.

The T-DNA consists of the following regions :

(i) An *one* region, consisting of three genes (two genes *tms/1* and *tms/2* representing 'shooty locus' and one gene *tmr* representing 'rooty locus'), is responsible for the biosynthesis of two phytohormones, namely indole-acetic acid or IAA (an auxin) and isopentyladenosine 5'-monophosphate (a cytokinin). These genes encode the enzymes responsible for the synthesis of these phytohormones, so that the incorporation of these genes in plant nuclear genome leads to the synthesis of these phytohormones in the host plant. The phytohormones in their turn alter the developmental programme, leading to the formation of crown gall.

(ii) An *os* region is responsible for the synthesis of unusual amino acid or sugar derivatives, which are collectively given the name opines. Opines are derived from a variety of compounds (e.g. arginine + pyruvate), that are found in plant cells. Two most common opines are octopine and nopaline. For the synthesis of octopine and nopaline, the corresponding enzymes octopine synthase and nopaline synthase are coded by T-DNA. Depending upon whether the Ti plasmid encodes octopine or nopaline, it is described as octopine-type Ti plasmid or nopaline-type Ti plasmid. Many organisms including higher plants are incapable of utilizing opines, which can be effectively utilized by *Agrobacterium*. Outside the T-DNA region, Ti plasmid carries genes that catabolize the opines, which are utilized as a source of carbon and nitrogen.

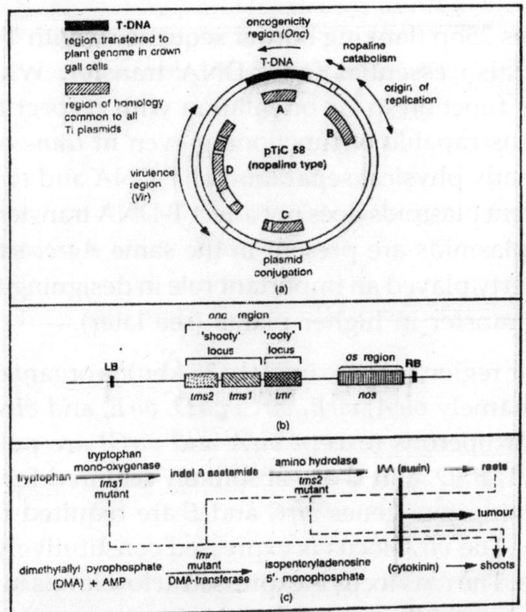

(a) General features of Ti plasmids; (b) general structure of the T-DNA; (c) phytohormone biosynthesis in crown gall tissues (Redrawn from D. Grierson 'Plant Genetic Engineering' 1991).

The T-DNA regions on all Ti and Ri plasmids are flanked by almost perfect 25bp direct repeat sequences, which are essential for T-DNA transfer, acting only in *cis* orientation. It has also been shown that any DNA sequence, flanked by these 25bp repeat sequences in the correct orientation, can be transferred to plant cells, an attribute that has been successfully utilized for *Agrobacterium* mediated gene transfer in higher plants leading to the production of transgenic plants.

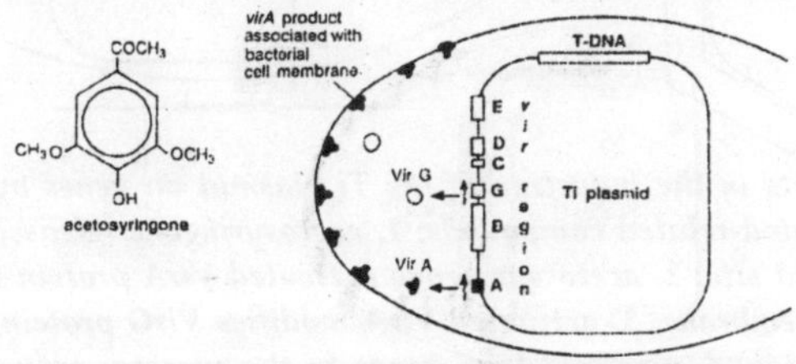

Ti plasmid gene expression in vegetative Agrobacterium prior to vir gene induction by the plant phenolic compound acetosyringone.

Besides 25bp flanking border sequences (with T-DNA), *vir* region is also essential for T-DNA transfer. While border sequences function in *cis* orientation with respect to T-DNA, *vir* region is capable of functioning even in *trans* orientation. Consequently physical separation of T-DNA and *vir* region on two different plasmids does not affect T-DNA transfer, provided both the plasmids are present in the same *Agrobacterium* cell. This property played an important role in designing the vectors for gene transfer in higher plants (see later).

The *vir* region (approximately 35 kbp) is organized into six operons, namely *virA, virB, virC, virD, virE,* and *virG* (Fig.), of which four operons (except *virA* and *virG)* are polycistronic. Genes *virA, B, D,* and *G* are absolutely required for virulence; the remaining two genes *vir*C and *E* are required for tumour formation. The *virA* locus is expressed constitutively under all conditions. The *virG* locus is expressed at low levels in vegetative cells, but is rapidly induced to higher expression levels by exudates from wounded plant tissue. The *virA* and *virG* gene products regulate the expression of other *vir* loci (Fig.).

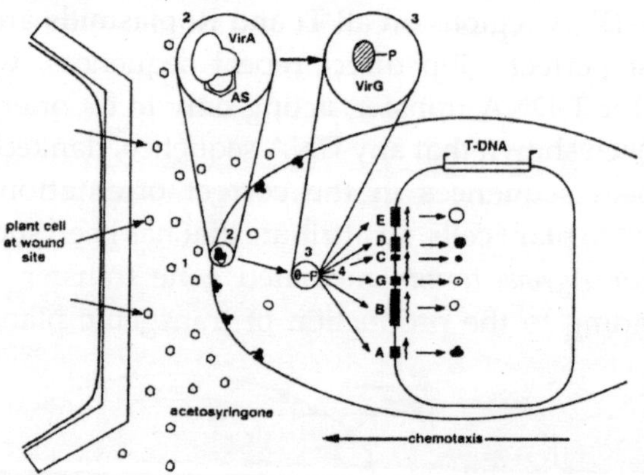

Key events in the induction of the Ti plasmid vir genes by acetoyringone and other wound-related compounds; 1. acetosyringone released by plant cells at the wound site; 2. aceto syringone activated VirA protein located in the bacterial membrane; 3. activated VirA modifies VirG protein, possibly by phosphorylation; 4. modified VirG binds to the operator sequences of the vir genes resulting in their activation or up-regulation (Redrawn from D. Grierson, 'Plant Genetic Engineering', 1991).

The *virA* product (VirA) is located on the inner membrane of *Agrobacterium* cells and is probably a chemoreceptor, which senses the presence of phenolic compounds (found in exudates of wounded plant tissue), such as acetosyringone and β-hybroxyaceto-syringone. Signal transduction proceeds via activation (possibly phosphorylation) of VirG (product of gene *virG),* which in its turn induces expression of other *vir* genes.

T-DNA Transfer Process : An early event in the T-DNA transfer process is the nicking of T-DNA of the Ti plasmid at two specific sites, each between the third and fourth base of the bottom strand of each 25bp repeat. This initiates DNA synthesis from the nick in the right hand 25bp repeat sequence in 5' — 3' direction thus displacing a single T-DNA strand.

This T-DNA single strand forms a complex with protein VirE and gets transported to the plant nucleus (Fig.). The *virD* operon encodes an endonuclease that produces the nicks in the border sequences. Several gene products of the *virB* operon have been identified in the bacterial envelope, a location, which suggests that they may play a role in directing T-DNA transfer extracellularly. The functions of several other vir gene products are largely unknown.

Apart from the role of Ti plasmid, the genes located on *Agrobacterium* chromosome also help in virulence. These genes are involved in the synthesis and secretion of glucons, cellulose fibrils and cell surface proteins. These loci are constitutively expressed and are also found in other soil bacteria associated with higher plants. Thus, these loci play a more general role in the virulence of *Agrobacterium,* and thus also in the *Agrobacterium* mediated gene transfer.

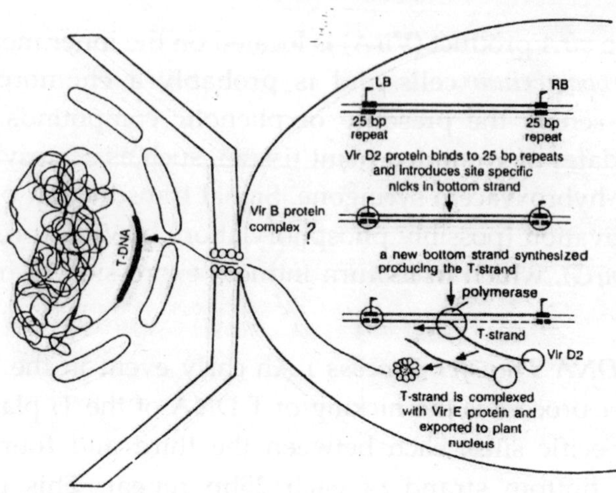

Molecular events in the transfer of the T-DNA to plant cells (Redrawn from D. Grierson 'Plant Genetic Engineering, 1991).

Vectors based on Ti and Ri plasmids : The following properties of Ti plasmids did not allow their direct use as vectors :

(i) large size,

(ii) absence of unique restriction enzyme sites and

(iii) tumour induction property.

Therefore, only the useful attributes of Ti plasmid have been used in designing several plant transformation vectors. Since tumour cells can not develop into normal shoots, disarming of Ti plasmids was an essential step in designing useful vectors. This was achieved by replacement of tumour inducting genes in T-DNA (T-DNA is found in both Ti and Ri plasmids) by selectable markers (like *nptII)*, providing resistance against antibiotics like kanamycin. Promoters and polyadenylation signals isolated from octopine or nopaline synthase genes were used for expression of selectable markers. Other powerful promoters included CaMV35S and CaMV19S isolated from cauliflower mosaic virus.

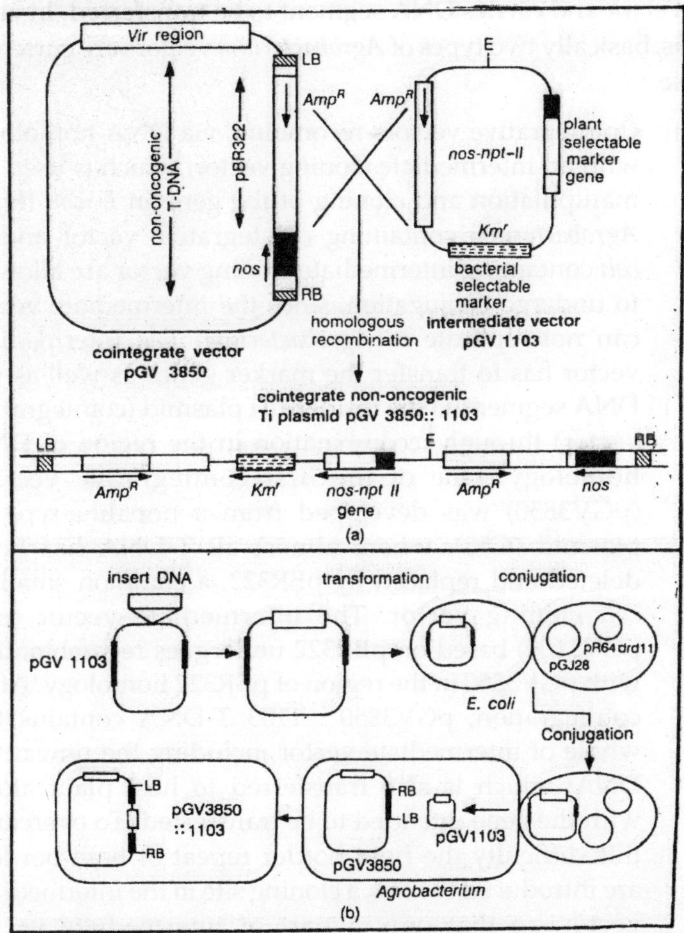

(a) Use of the cointegrate vector pGV3850 with the intermediate pGV1103. (b) Preparation of a cointegrative vector for the transfer of a desired DNA segment. (Redrawn from D. Grieson 'Plant Genetic Engineering, 1991).

As is obvious from the discussion in the previous section, T-DNA (disarmed) and *vir* genes are two essential elements for designing transformation vectors. While T-DNA with border sequences allows manipulation of DNA sequences intended to be transferred, *vir* genes allow infection of host plant. Since *vir* genes can function even in trans configuration, it is not essential that *vir* genes be present on the same plasmid, which functions as a vector and carries DNA segment to be transferred. In view of this, basically two types of *Agrobacterium* vectors are currently in use :

(i) Cointegrative vectors recombine, via DNA homology, with an intermediate cloning vector, which is used for manipulation and cloning of the gene in *E. coli* (Fig.). *Agrobacterium* containing cointegrative vector and *E. coli* containing intermediate cloning vector are allowed to undergo conjugation, since the intermediate vector can not replicate in *Agrobacterium.* The intermediate vector has to transfer the marker genes as well as the DNA segment to the resident Ti plasmid (cointegrative vector) through recombination in the region of DNA homology. One of the first cointegrative vectors (pGV3850) was developed from a nopaline-type Ti plasmid (C58), where almost all T-DNA has been deleted and replaced by pBR322, a common small *E. coli* cloning vector. The intermediate vector (e.g. pGV1103) based on pBR322 undergoes recombination with pGV3850 in the region of pBR322 homology. After cointegration, pGV3850 : 1103 T-DNA contains the whole of intermediate vector including the unwanted DNA, which is also transferred to host plant along with the gene intended to be transferred. To overcome this difficulty the right border repeat or both borders are introduced to flank a cloning site in the intermediate vector, so that only a part of intermediate vector cointegrates. The utility of this vector system is however limited.

(ii) Binary vectors are based on the principle that *vir* genes may be located on a *'helper'* Ti plasmid having the whole of T-DNA deleted (e.g. pAL4404), because *vir* genes can function even in *trans* configuration. In this case, T-DNA is found on a separate vector (binary vector) designed to replicate in both *E. coli* and *Agrobacterium* and capable of conjugal transfer between these two bacterial species.

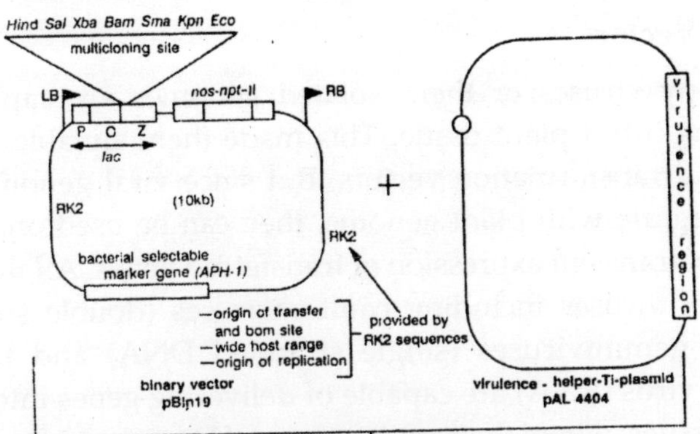

Typical binary vector strategy (Redrawn from D. Grierson, 'Plant Genetic Engineering, 1991).

Many binary vectors have been developed, which differ in size and the source of 25bp repeat sequences, plant selection marker, bacterial selection marker and cloning sites. pBinl9, which was designed in 1984, is still popular and is based on wide host-range replicon pRK252. This vector contains the following elements : (i) kanamycin resistance gene *(APH-I)* for selection in bacteria, (ii) T-DNA borders derived from pTiT37, (iii) a plant-selectable transformation marker *(nptII)* isolated from transposon Tn5 (this marker is associated with promoter and polyadenylation signal derived from nopaline synthesis gene or *nos)*, (iv) a multiple cloning site derived from pUC19 and housed within *lacZ* (β -galactosidase gene) region. Bacterial colonies containing pBinl9 are recognized by loss of blue colour on IPTG/X-GAL plates.

Unlike the cointegrative vectors, binary vectors need not have any homology with resident Ti plasmid and are capable of autonomous replication. Binary vectors based on Ri plasmid of *A. rhizogenes* have also been prepared. One such binary vector (pARS8) in a virulent strain (A4) of *A. rhizogenes* was used for the production of transgenic plants in tomato. Other examples of vectors based on Ri plasmids are also available in literature.

Virus Vectors

Many viruses or their isolated genomes are capable of infecting intact plant tissue. This made them suitable for use as plant transformation vectors. But since viral genomes can not integrate with plant genome, they can be used only for a study of transient expression of transferred genes. All the three kinds of viruses including caulimoviruses (double stranded DNA), geminiviruses (single stranded DNA) and tobacco mosaic virus (RNA) are capable of delivering genes into intact plant tissues, where they are expressed. However, virus vectors can not be used for the production of stably transformed plants, popularly described as transgenic plants.

Transformation Techniques using Agrobacterium : *Agrobacterium* infection (utilizing its plasmids as vectors) has been extensively utilized for transfer of foreign DNA into a number of dicotyledonous species and also into a few monocotyledonous plants. The success in this approach for gene transfer has resulted from improvement in tissue culture technology. However, initially monocotyledons could not be successfully utilized for *Agrobacterium* mediated gene transfer except a solitary example of *Asparagus.* The reasons for this are not fully understood, because T-DNA transfer does occur at the cellular level. It is possible that the failure in monocots lies in the lack of wound response of monocotyledonous cells. However, more recently (1990-2000) successful *Agrobacterium* mediated transformation has been achieved in several monocots including maize, wheat, rice, banana, etc.

Prerequisites for Production of Transgenic Plants : The important requirements for *Agrobacterium* mediated gene transfer in higher plants include the following :

(i) The plant explants must produce acetosyringone or other active compounds in order to induce *vir* genes for virulence; alternatively *Agrobacterium* may be preinduced with synthetic acetosyringone.

(ii) Induced agrobacteria should have access to cells that are competent for transformation; for gene transfer to occur, cells must be replicating DNA or undergoing mitosis (wounded and dedifferentiated cells, fresh explants or protoplasts have these properties).

(iii) Often transformed tissues or explants do not regenerate and it is difficult to combine transformation competence with totipotency (regeneration ability); therefore, the transformation competent cells should be able to regenerate in whole plants, a combination that can be easily achieved only in some species, such as tobacco. In some cases, undifferentiated cells of embryos may undergo transformation, so that the embryos may develop into chimeric plants.

Explants used for Transformation

The explants used for inoculation or cocultivation with *Agrobacterium* carrying the vector, include protoplasts, suspension cultured cells, callus cell clumps (undifferentiated and proembryogenic), thin cell layers (epidermis), tissue slices, whole organ sections (e.g. leaf discs, sections of roots, stems or floral tissues), etc. Wounding and inoculation of whole plants may also be used.

Marker Genes for Selection and Scoring of Transformed Cells/Calli or Shoots : After explants are inoculated with *Agrobacterium* carrying the requisite vector having the gene of interest, we need to select the transformed cells/tissues. This is facilitated by the presence of selectable marker genes available in the vector (see Table). The selectable marker genes enable the transformed cells to survive in media containing toxic levels of the selection agent, which is usually an antibiotic or a herbicide. Tobacco is used as a model transformation system, where explants (e.g. leaf discs) are placed on regeneration medium containing an antibiotic like kanamycin and transformed shoots can be obtained directly. Any cells, which are not transformed, die due to the presence of kanamycin. Other antibiotics and herbicides may require more judicious use, since even low concentrations can cause rapid cell death. In some cases, selection is exercised only after the regeneration is achieved, because adventitious root formation is sensitive to antibiotics.

Following successful selection, the next requirement is the regeneration of shoots from transformed calli. Since the explants may be heterogeneous and non-transformed tissue may escape initial rejection, several measures are used to ensure that very few non-transformed shoots escape the rejection. The use of leaf discs ensures this, because most cells at the wounded edges, which, undergo regeneration, are in contact with the selection agent.

Screening techniques also offer an alternative in which no selection pressure is imposed on cells or shoots developing from inoculated explants. In this case, samples of tissues are taken from all regenerated shoots and tested for the expression of a marker gene. A number of marker genes, described as screenable genes or scoreable genes or reporter genes are listed in Table. The most commonly used reporter genes are *nptII, cat, gus, lux,* and *pmi,* which will be briefly described.

Neomycin Phosphotransferase Gene (nptII) : This gene is used both as a selectable and a scoreable marker in experiments involving transfer of genes leading to the production of transgenic plants. It imparts kanamycin resistance, so that the transformed tissue can be selected on kanamycin. An assay for NPTII enzyme is also used to detect its presence in transformed tissue or transgenic plants. The gene for NPTII enzyme is often used with *nos* promoter, which drives its synthesis. In some cases, *nptII* gene had adverse effect on the expression of the desirable gene introduced (eg. *bt2* gene for insect resistance; see next chapter), so that alternative approaches for improving its expression had to be used.

Reporter Genes Used as Scoreable Markers, and their Enzyme Assays.

Reporter gene for	*Substrate and assay*	*Identification*
1. Chloramphenicol acetyl-transferase (CAT)	(^{14}C chloramphenicol + acetyl coA; TLC separation	Detection of acetyl chloramphenicol by autoradiography
2. Neomycin phosphotransferase (NPTII)	Kanamycin + (^{32}P) ATP (in situ assay)	Radioactivity detection (also dot blots)
3. β -glucuronidase (GUS; *uidA* gene)	Glucuronides (PNPG; x-GLUC, NAG, REG)	Fluorescence detection: coloriraetric, fluorometric and histochemical
4. (β -galactosidase (lac Z)	β -galactoside (X-gal)	Colour of the colony
5. Luciferase	(a) Decanal and $FMNH_2$ (b) ATP + O_2 + luciferin	Bioluminiscence (exposure of x-rays film)
6. Octopine synthase (OCS)	Ariginine + pyruvate + NADH	Electrophoresis
7. Nopaline synthase (NOS)	Arginine + ketoglutaric acid + NADH	
8. Green fluorescent protein (GFP)	No substrate is needed	Emission of green light when exposed to UV light

PNPG = p-nitrophenyl glucuronide ; x-GLUC = 5-bromo, 4-chloro, 3-indolyl glucuronide;
NAG = naphthol AS-BI glucuronide ; REG = resorufin glucuronide.

For an enzyme assay of NPTII, the enzyme is first fractionated using non-denaturing polyacrylamide gel electrophoresis (PAGE). Since the enzyme detoxifies kanamycin by phosphorylation, radioactively labelled ATP (P^{32}) is used with kanamycin in an agar layer, which is used to cover the gel containing the enzyme. The whole set is incubated at 35°C and the phosphorylation leading to incorporation of ^{32}P in kanamycin can be detected by autoradiography. As indicated in Table, a more convenient and easier enzyme assay can be conducted using dot blots of cell extracts (no DNA or enzyme extraction is necessary) obtained on a filter. The filter with dot blots is incubated with the substrate (kanamycin) and is then subjected to autoradiography to detect the presence of NPTII enzyme.

Chloramphenicol Acetyl Transferase Gene (cat) : This gene is primarily used as a reporter gene or a scoreale marker and not as a selectable marker. The gene first isolated from *E. coli* codes for an enzyme chloramphenicol acetyl transferase (CAT), which is absent in mammals and higher plants, so that whenever the gene *cat* is transferred with a gene construct, its presence can be detected by enzyme assay. Rarely screening for this enzyme may also be used for selection of transformed regenerants, although no selection pressure can be applied. The enzyme uses acetyl CoA + chloramphenicol (^{32}P) as substates and helps in the transfer of acetyl group to chloramphenicol. The presence of acetyl chloramphenicol is detected through autoradiography. The gene *cat* has also been used for identification of a number of regulatory sequences.

β-glucuronidase (GUS) gene *(uidA)*. The enzyme β-glucuronidase, popularly described as GUS, breaks down glucuronides giving a coloured reaction, so that its presence can be detected *in situ*, i.e. inside the plant tissue, used either as thin sections or in any other form. Several glucuronides, which can be used as substrates, include p-nitro phenyl glucuronide (PNPG), 5-bromo, 4-chloro, 3-indolyl glucuronide (BCIG), naphthol AS-B1 glucuronide (NAG) and resorufin glucuronide (REG). The enzyme GUS is coded by a gene *uidA*, first isolated from *E. coli*. The major advantage of using this reporter gene (scoreable marker) lies in its assay, which requires no DNA extraction, electrophoresis or autoradiography.

Luciferase gene (lux) : In the presence of a suitable substrate, the enzyme luciferase confers on the organism carrying it, the ability to glow in the dark. In other words the organism carrying this enzyme exhibits luminiscence on an assay for the enzyme luciferase. The enzyme luciferase differs in bacteria and firefly. While in bacteria, it consists of two peptide subunits coded by genes *luxA* and *luxB*, in firefly it consists of a single polypeptide coded by a gene *lux*. The enzyme luciferase in these two systems uses different substrates for luminiscence. Therefore, depending upon the source of luciferase gene used as a scoreable marker gene, the assay for detection of its presence also differs. When the source of the gene is bacteria, the substrate used is an aldehyde (decanal), which is supplied exogenously with $FMNH_2$. Alternatively, when the source is firefly, luciferin (a heterocyclic carboxylic acid) is instead used as a substrate, and is supplied with ATP and oxygen (Table).

Green fluorescent protein gene. The gene coding for green fluorescent protein (GFP), was isolated in the year 1992 from the jelly fish *Aequorea victoria*. This gene when used for production of transgenic plants, allow non-destructive, real time, *in vivo* assay. It has been introduced in bacteria, nematodes, fruitfly *(Drosophila)*, mice and several plant species. GFP is a 27 kD monomer, that has the unique characteristic of emitting green light when exposed to ultraviolet or UV light (~ 395 nm) or blue light (~490 nm). This gene differs from other reporter genes, since no substrate or enzyme is needed for detection. Therefore, this will prove to be a universal transgenic marker, because it is species independent. Any plant and its progeny expressing GFP can be visually tracked in real time, so that the plants could be easily monitored. When GFP gene is under the control of a constitutive promoter, the leaves fluoresce green when exposed to UV or blue light and viewed under low ambient light, giving contrast to the pinkish hue emitted by non-transgenic plants.

The gene for GFP has been mutagenized, resynthesized for higher expression in plants. These modifications involved altered codons without changing the amino acids coded (codon usage for optimum expression in higher plants). This modified GFP gene gives stable and improved expression in higher plants. This also allows testing of many plants in the same field and will permit detection of any chance escape of the gene to other species growing in the vicinity, which can then be monitored or destroyed as needed.

Phosphomannose Isomerase Gene (pmi) : Although the genes for antibiotic and herbicide resistance as selectable markers for plant transformation have been shown to be completely safe and seem to pose no risk to the farmers or consumers, public concerns have been voiced against these markers, particularly when used in food crops. Therefore, efforts have been made to use systems (e.g., *cre-lox;* see later), which will remove the selectable marker, after the selection of transformed cells is over. However, to address public concerns, efforts are also being made to phase out the use of antibiotic resistance genes as markers. As a step in this direction, during 2000-2001, a company named Novartis developed and used a selectable marker system that makes use of phosphomannose isomerase gene *(pmi)* derived from *E. coli.* The product of this gene, the enzyme phosphomannose isomerase (PMI) converts mannopse-6-phosphate (obtained from mannose due to action of hexokinase) to fructose-6- phosphate, which can be utilized by transformed cultured cells. The method for the use of this marker system in crop plants involves cells or tissues using any method, followed by placement of these cells in a culture medium, to which mannose is added as a substrate either as a sole carbon source or in combination with sucrose. Mannose has no adverse effect on plant cells, but selection occurs due to its phosphorylation to mannose-6-phosphate by hexokinase. In tissues lacking PMI, mannose-6-phosphate accumulates and the cells stop growing, thus offering selection of transformed cells, which can metabolise mannose-6-phospote due to the presence of gene *pmi*.

Agrobacterium-mediated Transformation in Monocots : Although, during early 1980s, transformation mediated by *Agrobacterium tumefaciens* had been successfully achieved in many dicotyledonous crops, initially it could not be used in monocotyledonous crops like maize, wheat, rice, barley, oats, etc. In these monocots, biolistics and direct gene transfer to protoplasts by electroporation and polyethyleneglycol (PEG) mediated uptake were often utilized successfully. However, *Agrobacterium-mediated* transformation was later (in 1990s) successfully achieved in several cereal crops including maize, rice, wheat and barley. *Agrobacterium* mediated gene transfer was preferred over other methods of gene transfer, since it offered a number of advantages including the following: (i) defined transgene integration; (ii) integration of low copy number; (iii) preferential integration into transcriptionally active region. In view of this, concerted efforts were made and success was achieved in *Agrobacterium* mediated transfer of genes in monocots, which was increasingly utilized during 1994-2000.

The apparent difficulties encountered in *Agrobacterium* mediated transformation initially led many workers to believe that this approach had very little potential in cereals. The success in using this approach in cereals has been ascribed to a combination of factors including the following : (i) Choice of a supervirulent *Agrobacterium* strain, (ii) use of a suitable target material (e.g. embryo) undergoing active cell divisions and (iii) use of better promoter like that from maize ubiquitin gene in plasmid pDM805, to drive the expression of the selectable marker gene.

Agroinfection and Gene Transfer

Agroinfection is a phenomenon, in which a virus infects a host as a part of T-DNA of Ti plasmid carried by *Agrobacterium.* Viral DNA can be integrated into the T-DNA and can be delivered into plant cells with the normal *Agrobacterium* T-DNA transfer process. After infection viral DNA is released to form functional virus that replicates and spreads systemically. Agroinfection may also lead to the integration of viral DNA so that transgenic plant containing integrated viral DNA can be produced.

In maize, agroinfection with maize streak virus has been demonstrated. This suggested that *Agrobacterium* based vector system can be used for genetic engineering in cereals, although ordinarily *Agrobacterium* does not infect monocotyledons. Thus agroinfection can lead to the production of transgenic plants, even though it has no better chances of yielding transgenic cereals than does *Agrobacterium* infection alone. However, agroinfection has great potential for studies in virus biology, because it can transfer deletion mutations or even single viral genes.

Physical Delivery Methods or DNA Mediated Gene Transfer (DMGT) : *Agrobacterium* mediated gene transfer has been the most commonly used method of gene transfer in plants but the cereals, comprising the most important food crops, were initially not amenable to this method of gene transfer. Further, in many crops including cereals and legumes, initially the tissue culture techniques for regeneration were also not very successful. These two limitations forced intensive search for alternative methods for gene transfer. Physical delivery of DNA or DNA mediated gene transfer (DMGT), as it is often described, emplyos methods, which can be grouped according to the type of target cell. For instance, chemically stimulated endocytosis of plasmids or that of DNA-loaded liposomes, and electroporation were employed for delivery of DNA to protoplasts only.

On the other hand, techniques like microinjection, macroinjection and shooting with microprojectiles could be used with a variety of explants (e.g. immature embryos, organ meristems, gametes, zygotes, etc.). These latter techniques achieved significance, because regeneration of transformed plants from protoplast derived tissues was initially difficult in many species particularly in cereals (although encouraging results were later obtained in case of maize, rice, wheat, oats, etc.).

Chemically Stimulated DNA Uptake by Protoplasts : Direct DNA uptake by protoplasts can be stimulated by chemicals like polyethylene glycol (PEG). The technique is so efficient that virtually every protoplast system has proved transformable. PEG is also used to stimulate the uptake of liposomes and to improve the efficiency of electroporation (see below for details). PEG at high concentration (15-25%) will precipitate ionic macromolecules such as DNA and stimulate their uptake by endocytosis without any gross damage to protoplasts. This is followed by cell wall formation and initiation of cell division. These cells can now be plated at low density on selection medium.

Initial studies using the above method were restricted to *Petunia* and *Necotiana*. However, other plant systems (rice, maize, etc.) were also successfully used later. In these methods, PEG was used in combination with pure Ti plasmid, or calcium phosphate precipitated Ti plasmid mixed with a carrier DNA. Transformation frequencies upto 1 in 100 have been achieved by this method. Nevertheless, there are serious problems in using this method for getting transgenic plants, mainly due to difficulties encountered in plant regeneration from protoplasts.

Microinjection and Macroinjection

Plant regeneration from transformed protoplasts, still remains a problem. Therefore cultured tissues, that encourage continued development of immature structures, provide alternate cellular targets for transformation. These immature structures may include immature embryos, meristems, immature pollen, isolated ovules, embryogenic suspension, cultured cells, etc. The main disadvantage of this technique is the production of chimeric plants with only a part of the plant transformed. However, from this chimeric plant, transformed plants of single cell origin can be subsequently obtained. Utilizing this approach, transgenic chimeras were actually obtained in several crops including oilseed rape *(Brassica napus).*

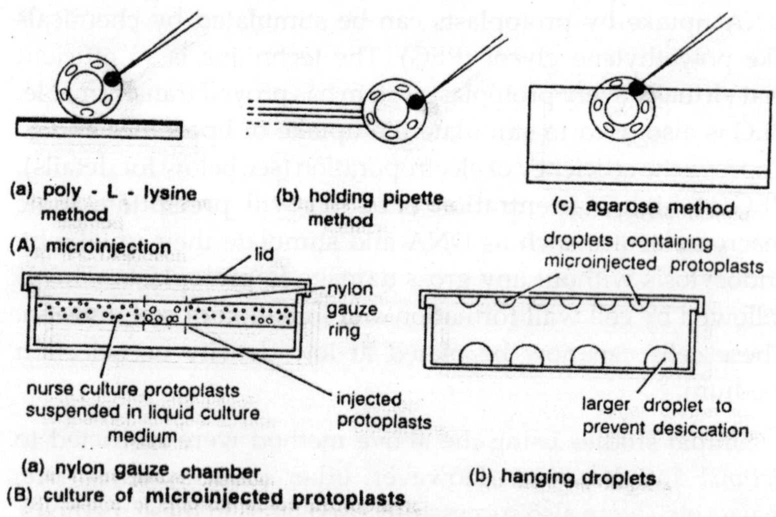

Methods for microinjection and culture of microinjected plant protoplast.

When cells or protoplasts are used as targets in the technique of microinjection, glass micropipettes each with 0.5-10 μ m diameter tip are used for transfer of macromolecules into the cytoplasm or the nucleus of a recipient cell or protoplast. The recipient cells are immobilized on a solid support (cover slip or slide, etc.) or artificially bound to a substrate or held by a pipette (Fig.) under suction (as done in animal systems). Often a specially designed micromanipulator is employed for microinjecting the DNA. Although, this technique gives high rate of success, the process is slow, expensive and requires highly skilled and experienced personnel.

DNA macroinjection employing needles each with diameter greater than cell diameter, has also been tried. In rye *(Secale cereale)*, a marker gene was macroinjected into the stem below the immature floral meristem, so as to reach the sporogenous tissue (De la Pena et al., 1987) leading to successful production of transgenic plants. Unfortunately, this technique could not be successfully repeated with any other cereal, when tried in several laboratories. Therefore, doubt is expressed about the validity of earlier experiments conducted with rye (Potrykus, 1991).

Microprojectiles (biolistics or particle gun) for Gene Transfer : During early 1990s, it was shown that DNA delivery to plant cells is also possible, when heavy microparticles (tungsten or gold) coated with the DNA of interest are accelerated to a very high initial velocity (1,400 ft per sec). These microprojectiles, each normally 1-3 μ m in diameter, are carried by a 'macroprojectile' or the 'bullet' and are accelerated into living plant cells (target cells can be pollen, cultured cells, cells in differentiated tissues and meristems) so that they can penetrate cell walls of intact tissue. The acceleration is achieved either by an explosive charge (cordite explosion) or by using shock waves initiated by a high-voltage electric discharge. The designs of two particle guns used for acceleration of microprojectiles are shown in Figure.

Transgenic plants using the above technique have been obtained in many cases including soyabean, tobacco, maize, rice, wheat, etc. Transient expression of genes transferred in cells by this method has also been observed in onion, maize, rice and wheat. During 1990s, there was no other gene transfer approach, which met with so much of enthusiasm. Consequently considerable investment was also made in experimentation and manpower for development of this technique. The advantages of this method over microinjection include the following :

(i) thousands of particles are accelerated at the same time, causing multiple hits resulting in transfer of genes into many cells simultaneously;

(ii) since intact cells can be used, some of the difficulties encountered with the use of protoplasts are automatically circumvented;

(iii) the method is universal in its application, so that cell type, size and shape or the presence/absence of cell walls do not significantly alter its effectiveness.

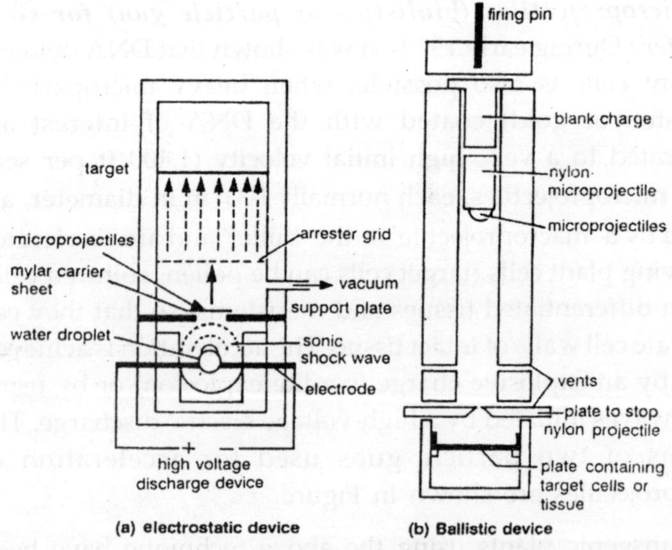

Microprojectile acceleration devices.

In view of this, particle bombardment method using microprojectiles was used in a variety of plant species, particularly the cereals. The method was also used successfully for transfer of genes into the plastid genome. However, the emphasis on the use of particle gun decreased towards the end of 20th century, due to the success achieved in cereals with *Agrobacterium* mediated gene transfer.

Electroporation Method for Gene Transfer

This method is based on the use of short electrical pulses of high field strength. These pulses increase the permeability of protoplast membrane and facilitates entry of DNA molecules into the cells, provided the DNA is in direct contact with the membrane. In view of this, for delivery of DNA to protoplasts, electroporation is one of the several routine techniques for efficient transformation. However, since regeneration from protoplasts is not always possible, cultured cells or tissue explants are often used. Consequently, it is important to test whether electroporation could transfer genes into walled cells. In most of these cases no proof of transformation was available.

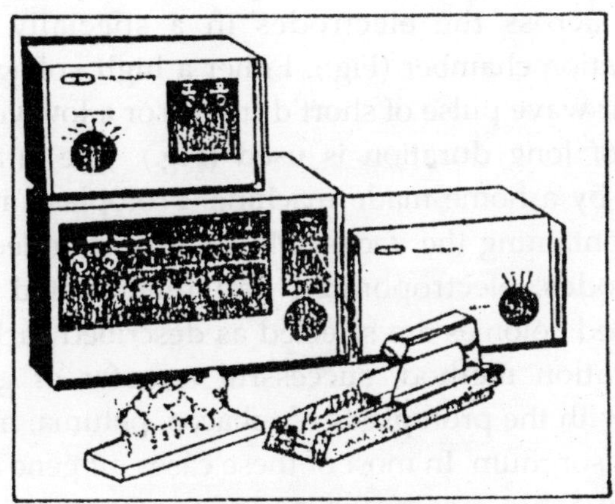

An electroporation system, suitable for transformation of bacteria, mammalian cells and plant cells.

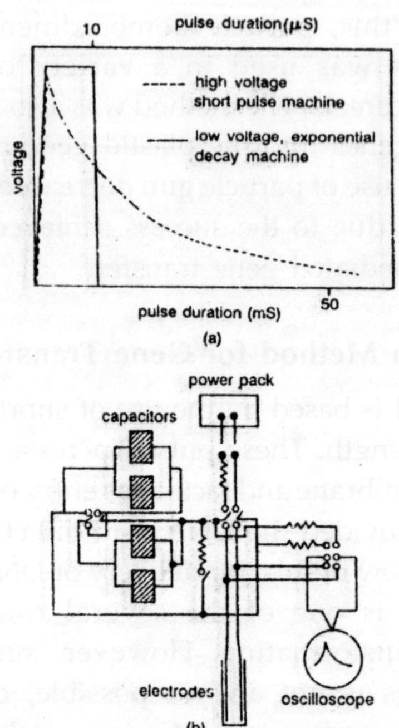

(a) Details of the output characteristics of the two principal types of electroporation devices in use: (b) circuit diagram of low voltage, exponential decay machine. (Redrawn from D. Grierson. 'Plant Genetic Engineering', 1991).

The electroporation pulse is generated by discharging a capacitor across the electrodes in a specially designed electroporation chamber (Fig.). Either a high voltage (1.5 kV) rectangular wave pulse of short duration or a low voltage (350 V) pulse of long duration is used (Fig.). The latter can be generated by a home made machine. Protoplasts in an ionic solution containing the vector DNA, are suspended between the electrodes, electroporated and then plated as usual. Transformed colonies are selected as described earlier. Using electroporation method, successful transfer of genes was achieved with the protoplasts of tobacco, petunia, maize, rice, wheat and sorghum. In most of these cases *cat* gene associated with a suitable promoter sequence was transferred. Transformation frequencies can be further improved by (i) using field strength of 1.25 kV/cm, (ii) adding PEG after adding DNA, (iii) heat shocking protoplasts at 45°C for 5 minutes before adding DNA and (iv) by using linear instead of circular DNA.

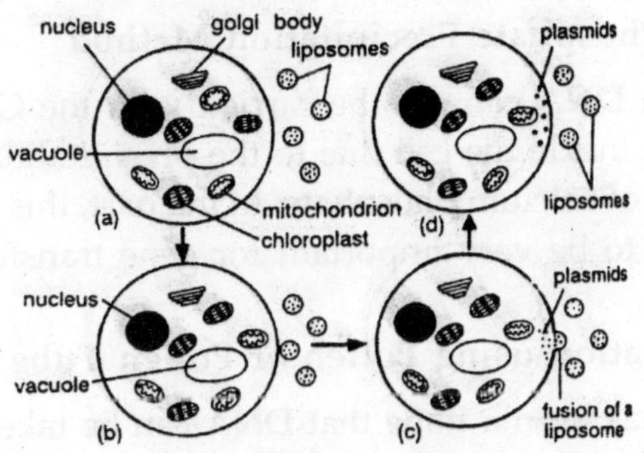

Fusion of plasmid-filled liposomes with isolated protoplasts as a method of plasmid uptake.

Liposome Mediated Gene Transfer

Liposomes are small lipid bags, in which large number of plasmids are enclosed. They can be induced to fuse with protoplasts using devices like PEG, and therefore have been used for gene transfer. The technique offers following advantages :

(i) protection of DNA/RNA from nuclease digestion,

(ii) low cell toxicity,

(iii) stability of nucleic acids due to encapsulation in liposomes,

(iv) high degree of reproducibility and

(v) applicability to a wide range of cell types. In this technique, DNA enters the protoplasts due to endocytosis of liposomes, involving the following steps:

   (i) adhesion of liposomes to the protoplast surface,

   (ii) fusion of liposomes at the site of adhesion and

(iii) release of plasmids inside the cell. The technique has been successfully used to deliver DNA into the protoplasts of a number of plant species (e.g. tobacco, petunia, carrot, etc.).

Calcium Phosphate Precipitation Method

Foreign DNA can also be carried with the Ca++ ions, to be released inside the cell due to the precipitation of calcium in the form of calcium phosphate. In the past, this method was considered to be very important for gene transfer in plants.

Transformation using Pollen or Pollen Tube

There has been a hope that DNA can be taken up by the germinating pollen and can either integrate into sperm nuclei or reach the zygote through the pollen tube pathway. Both these approaches have been tried and interesting phenotypic alterations suggesting gene transfer have been obtained. In no case, however, unequivocal proof of gene transfer has been available. In a number of experiments, when marker genes were used for transfer, only negative results were obtained. Several problems exist in this method and these include the presence of cell wall, nucleases, heterochromatic state of acceptor DNA, callose plugs in pollen tube, etc. Transgenic plants have never been recovered using this approach and this method, though very attractive, seems to have little potential for gene transfer.

Incubation of Dry Seeds, Embryos, Tissues or Cells in DNA: Incubation of dry seeds, embryos, tissues or cells in known DNA (viral or non-viral DNA having defined marker genes) has been tried in many cases and expression of defined genes has been witnessed. However, in no case proof of integrative transformation could be available. In all these cases, plant cell walls not only work as efficient barriers, but are also efficient traps for DNA molecules. It would be very surprising if DNA can cross cell walls efficiently without permeabilizing them either by PEG, or by electroporation or by any other device.

Transformation by Ultrasonication

In wheat, tobacco and sugarbeet, explants after being cultured for a few days, were sonicated with plasmid DNA (carrying marker genes like *cat, nptII* and *gus*). When sonicated calli were transferred to selective medium, shoots were obtained, although all control calli (not sonicated with plasmid DNA) died. In tobacco, transgenic plants were obtained at a frequency of 22 per cent. This work was conducted at the Biotechnology Research Center at Beijing (China).

Removal of Selectable Marker Genes from Transgenic Plants

Need for Removal of Selectable Marker Gene : Removal of selectable marker genes from transgenic plants is desirable for the following reasons :

(i) Consumer groups have expressed concern about the wide-spread distribution of markers imparting resistance against antibiotics and herbicides, in transgenic foods; this poses theoretical risk of horizontal transfer of the marker genes to gut bacteria. It is argued that the resulting resistant bacteria will then not respond to antibiotics used as medicine.

(ii) Crosses between transgenic crops will increase the number of copies of the marker genes in a crop, thus not only increasing the probability of horizontal gene transfer, but also enhancing the likelihood of homology dependent gene silencing, thus limiting the long-term use of transgenic crops.

(iii) The presence of a marker gene will not allow transfer of an additional gene using the same marker gene (sometimes described as pyramiding of genes),

(iv) The marker gene, like the herbicide resistance gene, may be transferred to weeds, due to long term use of herbicide resistant transgenic crops, thus causing ecological problems.

Genetic Systems used for Removal of Marker Gene : Several transposable element systems and site-specific recombination systems are available for the removal of marker gene. These systems involve the use of transposase or recombinase enzyme, which mediates the deletion of the marker gene. The action of the enzyme is facilitated by the target sequences, which flank the marker gene. Following are two examples of many systems that are now available for removal of the marker gene :

(i) *Cre-loxP* recombinase system involves the use of gene constructs, which will have *loxP* sequences (34bp) flanking the marker gene to be removed, so that when gene *Cre* (recombinase enzyme from bacteriophage P1) is expressed, site *(loxP)* specific recombination occurs, and *loxP* sequences along with the marker gene will be excised and removed.

(ii) In another strategy reported in the year 2000, *nptII* gene flanked by two 352bp phage attachment *(attP)* regions of lambda (λ) phage is used as a marker gene (*attP* sequence helps λ phage in attachment to the bacterial genome). This is followed by identification of somatic tissue with deletion events following intrachromosomal recombination (ICR). For this purpose a vector pattP-ICR was constructed, in which *attP* flanked the marker gene *nptII. A* transformation booster sequence (TBS) next to the left *attP* site enhanced homologous and illegitimate recombination. In this system intrachromosomal recombination (ICR) deletes a 5.9kb sequence and produces a transgenic plant lacking the marker gene.

Examples of North American Plant Transformation Facilities.

Institute	*Target Crops*	*Contact*
Cornell University	Tomato, algae, fungi	Karen Kindle klk9@cornell.edu
Iowa State University	Corn, soyabean	Kan Wang kanwang @ iastate.edu
Texas A&M University Crop Biotechnology Center	Cotton, rice, sorghum, banana	Keerti Rathore rathore@genome.tamu.edu.
Taxes A&M University Forest Biotechnology Group	Loblolly, Virginia pine	Jean Gould jean@rsgis.tamu.edu.
University of Georgia, University of Kentucky, Ohio State University Center for Soybean Tissue Culture & Genetic Engineering	Soyabean	Glenn Collins agrl23@ukcc.uky.edu.
University of Nebraska, Lincoln	Wheat, soybean	Paul Staswick pstaswick@crcvms.unl.edu.
University of Wisconsin at Madison Biotechnology Center — lant Biotechnology Laboratory	Alfalfa, potato	Sandra Austin-Phillips saustinp@facstaff.wisc.edu.
NRC of Canada—Plant Biotechnology Institute Transgenic Plant Center	Canola, wheat, pea	Kutty Kartha kkarth@pbi.nrc.ca

Plant Transformation Facilities (PTFs) : Since infrastructure and the expertise for plant transformation may not be available everywhere, the concept of establishing plant transformation facilities has been welcome by many. For a number of important crop species, including grain legumes, cereals and conifers, transformation is technically demanding, so that a plant transformation facility if available on payment (or through collaboration) may be utilized. Keeping in view the above, in USA, a number of academic and government institutions have decided to establish core plant transformation facilities (Table). One can utilize these facilities particularly if the objective is research only, although the ownership of the final transgenic material, if commercialized, will be governed by the relevant intellectual property rights or IPRs.

8

Microbial Genomics

Microbes are also used for a variety of useful purposes, and can be improved through genetic engineering. The harmful effects of microbes are also well known, since they cause a variety of diseases both in animals (including humans) and crop plants. Microbes also cause damage to stored food and feed, making it necessary to arrange adequate proper facilities to store our food and animal feed. While the efficiency of useful microbes can be improved through genetic engineering, microbes can also be used as factories for the synthesis of novel products through transfer of novel genes from other organisms or through transfer of newly designed synthetic genes. For instance, genetically engineered microorganisms now produce a range of pharmaceutical products including a variety of enzymes at a reduced cost.

Genetic engineering can also be used for mitigating the harmful effects of a number of microbes, if the underlying mechanism of the harm caused is known in each individual case. Genetically engineered microbes can also be used for increased protection and productivity of crops, for biodegradation of xenobiotics and toxic wastes (bioremediation), and for production of cheap fuels. Besides these uses of genetically engineered microbes, microbes are also being subjected to extensive sequencing of their whole genomes using tools of genetic engineering, so that 'microbial genomics' has also become an important area of research. Genomes of as many as about 500 microbes have already been fully sequenced, and there are reports almost every week, of the completion of whole genome sequences of new microbes. Thus genetic engineering of microbes and microbial genomics has become a thrust area of research in microbiology and biotechnology. In this chapter we will briefly discuss the progress made in this area of research using suitable examples.

Applications of Genetically Engineered Bacteria : A variety of micro-organisms, particularly bacteria, have been modified through the techniques of genetic engineering to meet specific needs. These genetically engineered bacteria, have found applications in areas including the following :

(i) production of cheese,

(ii) production of growth hormones and Pharmaceuticals,

(iii) crop production and protection through biological control of insects, fungal diseases, frost damage, etc.,

(iv) biodegradation of xenobiotics (waste from non-biological systems) and toxic waste transformations,

(v) production of fuels and chemicals including antibiotics, enzymes and diagnostics and

(vi) extraction of metals from ores. Application of this technology will ultimately release beneficial agents in the environment. It is hoped that if safely deployed, genetically modified bacteria should be able to provide significant benefits in the management or environment systems. Some examples of these genetically engineered bacteria will be briefly discussed in this section.

GEM for rBGH/rBST to Increase Cow Milk : The recombinant Bovine Growth Hormone (rBGH), also called Bovine Somatotropin (rBST), is a genetically engineered hormone manufactured by Monsanto (a multinational company in USA). This is one of the first and the most important examples, where genetically engineered microbes have been used for economical yield of a useful product. The bioengineered hormone is injected in the cows every other week to force the cows to produce more milk than their bodies normally would. rBGH is similar, although not identical, to a hormone that the cow naturally produces. Increased level of this hormone boosts milk production, which made it initially popular. However, ever since its introduction to the marketplace a couple of years ago, this sparked a nationwide controversy in USA. The hormone has been shown to cause a number of problems with the milk. Among these, the main problems included the following :

(i) raising levels of pus in the cows treated with rBGH(increased amount of pus makes cows more susceptible to udder infections),

(ii) antibiotic residues in the milk so produced and

(iii) a cancer-accelerating hormone called IGF-1 in the milk (IGF-1 is associated with breast, prostate, and colon cancers).

Cheese from Chymosins Derived from GM Yeast : In order to make cheese, an enzyme preparation called rennet is added to milk. Rennet, which is made from the lining of a calfs stomach, contains a specific enzyme, called chymosin. Chymosin causes particles of milk protein (casein) to clump together into a solid gel. When the gel is cut, a liquid containing some protein and the milk sugar, lactose, drains out. This is called whey. The solid part left behind, containing the nutritious elements such as milk protein and calcium, is called curd and this is what is used to make the cheese. The pieces of curd are pressed together and stored in temperature-controlled rooms. During this time the cheese matures and the enzymes in the rennet and starter culture act on the protein and fat. This produces the distinctive cheesy flavour.

Chymosin from other Sources : During 1960s the Food and Agriculture Organisation (FAO) of the United Nations predicted a severe shortage of calf rennet. It was anticipated that an increased demand for meat would lead to more calves being reared to maturity so that less rennet would be available. Also a growing number of people following vegetarian diets do not eat cheese made with calf rennet. As a result, over the last thirty years several substitutes for calf rennet have been made available, allowing the supply of enzymes to keep pace with cheese production and providing alternative sources for vegetarians. Today there are two major sources of chymosins for coagulating milk : from animals (veal calves, adult cows and pigs) and from different kinds of fungi. In addition there are now chymosins derived from genetically modified microbes. In the latter case, copies of the genes responsible for chymosin production are isolated from calf stomach cells and incorporated into the genetic material of yeast cells. These can be grown on an industrial scale and the chymosin isolated.

Chymosin from Genetically Modified Microbes : Chymosin obtained by genetic modification has been subjected to rigorous tests to ensure its purity. Biochemical, toxicological and microbiological tests have also been carried out. The advantages are obvious. Fewer calves are required for rennet production, and since yeast cells can be grown quickly and in huge quantities, fluctuating demand for chymosin can easily be satisfied. Chymosin produced by genetically modified microbes is very pure and thus more predictable in terms of activity than its traditional counterpart. Genetically modified chymosin has been approved in most European countries although its use differs from country to country. In the UK, the Vegetarian

Society has given cheese made with genetically modified chymosin its seal of approval.

GEMs for Recombinant Pharmaceuticals : A number of pharmaceuticals that were earlier obtained from human beings, or by sacrificing animals are now synthesized in microbes or cultured mammalian cells.

Insulin Production in Bacteria : Human insulin is involved in the breakdown of sugar in human body, so that any deficiency for insulin will cause accumulation of sugar thus causing the disease 'diabetes'. In the past, treatment of human diabetes was dependent on animal insulin, which could be obtained only through sacrificing animals, thus making it expensive. Animal insulin sometimes also caused adverse reactions in some patients. Using genetic engineering, the gene for insulin could be cloned in bacteria, so that correct form of insulin can now be obtained at reduced cost in quantities never before available.

Production of Human Growth Hormone in Bacteria : Individuals having deficiency for human growth hormone (HGH) suffer with hypopituitary dwarfism, and are advised to take this hormone exogenously to overcome the problem. Although mainly used to treat some forms of dwarfism, HGH also has the potential to treat wounds, burns, fractures, and other conditions that heal slowly.

Like insulin gene, HGH gene could also be synthesized using recombinant DNA technology, and could be cloned in bacteria, which could then be used for production of this hormone in bioreactors in large quantities at a reduced cost. During late 1980s, recombinant HGH was approved for commercial marketing under the name prototropin in USA and under the name somatonorm in UK. The manufacture of prototropin and somatonorm assumed significance, because the use of HGH derived from human pituitary glands was halted, due to the fear that it may transfer some other fatal disease like Creutz Feldt-Jakob Disease (CJD) from the donor to the patient.

Tissue Plasminogen Activator (tPA) : Tissue plasminogen activator (tPA) is used to dissolve blood clots, thus reducing the risk of a heart attack or stroke. The gene for this protein has been cloned and transferred to cultured cells and microbes, to obtain large quantities of tPA at a reduced cost. Clinical trials have shown that recombinant (genetically engineered) tPA is safe and effective in dissolving blood clots in patients with heart diseases and thrombotic disorders. Genetic manipulations have also been used for increased production of tPA through regulation at the post-translational level. The protein tPA is known to remain associated with GRP78 in endoplasmic reticulum, thus reducing its secretion (GRP78 is a glucose regulated stress response protein). Utilizing antisense RNA technology, synthesis of GRP78 could be reduced at the translation level, which led to increased secretion of tPA in cultured cells, which mimic microorganisms.

Crop Production and Protection

Several bacterial strains have been modified by introduction of foreign genes to control insects (by production of endotoxins), fungal diseases (by production of chitinases, which suppress fungal flora in the soil), or frost damage (through production of mutants for ice nucleation gene or *ice*$^+$ responsible for frost damage). Deleterious bacteria and fungi or their effects have been controlled in some cases by using a variety of approaches, which include the following :

(i) production of avirulent mutant of the pathogen and releasing it to compete with the virulent strains;

(ii) production of antibiotic producing strains of plant associated bacteria to control pathogen through antibiotic production or

(iii) production of bacterial strains that will degrade the toxin produced by the pathogen.

There are also other positive measures, where the N_2 fixing efficiency of *Rhizobia* can be enhanced by transfer of useful *nif* genes and other related genes involved in nodulation, nitrogenase activity and other related functions. Some examples of these genetically engineered bacteria for the possible increase in crop production and protection are listed in Table.

Genetically Engineered Bacteria in Crop Production and Crop Protection.

Bacterium	*Altered trait*	*Possible use*
1. *Clavibacter xyli*	Transfer of *B. thuringiensis* delta endotoxin gene	Control of corn ear worm
2. *Pseudomonas fluorescens*	Transfer of *Serratia marcescens* chitinase gene	Control of fungal disease
3. *P. fluorescens, P. syringae*	Deletion of *'ice'* gene	Control of frost damage
4. *P. fluorescens*	Addition of *lac ZY*	Assessment of movement of bacteria for biological control ('take all disease' of wheat)
5. *A. radiobacter*	Deletion of *'tra'* gene of Agrocin 84 plasmid	Biological control of crown gall
6. *Rhizobium melilotii*	Additional copies of *'nif* gene	Increased efficiency of N_2 fixation

A List of Genetically Engineered Bacteria Used for Degradation of Xenobiotics and Toxic Wastes.

Bacterium	*Substrate that can be degraded*
1. *Pseudomonas capacia*	2, 4, 5 - trichloro- phenoxyacetic acid
2. *P. putida* & other spp (also *E. coli)*	2, 2, 5 - dichloro-propionate; mono and dichloroaromatics
3. *Alcaligenes* sp.	Dichlorophenoxyacetic acid; mixed chlorophenols; 1, 4-dichlorobenzene
4. *Acinetobactor* sp.	4 - chlorobenzene

GEMs for Bioremediation

Bacteria can be modified genetically to develop catabolic pathways for degradation of xenobiotics (wastes from nonbiological systems) and other waste material. Bacterial genes for this purpose are isolated from bacterial strains (or their plasmids) found at waste sites. Most of these genes are available from gram-negative bacteria (e.g. *Pseudomonas* spp.), which are themselves not very efficient degraders, since multiple genes may sometimes be needed for efficient biodegradation. Therefore, for efficient biodegradation, efficient degraders have to be prepared through genetic engineering and they need to be established in the environment at a required density. Some of the genetically engineered bacterial strains for biodegradation of waste material are listed in above Table.

GEM for Detection of Polluting PAHs in the Soil : Polyaromatic hydrocarbons (PAHs), including compounds like naphthalene, phenanthrene and anthracene, is one of the main source of pollution (due to oil spills) and their detection in the polluted soil is one of major steps in bioremediation process. In USA, *Pseudomonas fluorescence* isolated from PAH contaminated soils, was genetically engineered with *lux* gene from *Vibrio fischeri* in such a manner that in the presence of PAHs, they will emit light, which can be detected using light sensing device, thus permitting detection of PAHs in the soil.

GEMs for Treating Oil Spills : Genetic engineering of pseudomonads for environment cleaning has been an important area of research. Several species of *Pseudomonas* have been genetically engineered, for degradation of oil spills and other waste material, both on land and water. Ananda Chakrabarty, an NRI in USA, developed the first genetically engineered organism (a pseudomonas strain) for cleaning the oil spills. This *Pseudomonas* strain was capable of degrading 2,4,5-trichlorphen Oxyacetic acid (2,4,5-T). The strain contained two plasmids, each providing a separate degradative pathway, and therefore, was claimed to be effective in treating oil spills. Several other microbes were genetically engineered for treatment of oil spills (Table).

GEM for Sequestering Heavy Metals : A new approach for bioremediation of polluted soils was suggested in a study reported in 2000. The approach involved engineering of microorganisms to enhance their ability to sequester heavy metals in the soil. In this approach lower. The toxic metal within the soil remains bound to the GEM, so that it is less likely to be taken up either by the underground part (roots) of the terrestrial plants, or by other plants or animals living in the soil. The enhanced ability to sequester heavy metals (e.g. cadmium) was achieved by transfer of a mouse gene, encoding metallothionein *Ralstonia eutropha* (a natural inhabitant of soil). Metallothionein in the GEM was expressed on the outer surface of the cells to help in sequestering of cadmium (Fig.).

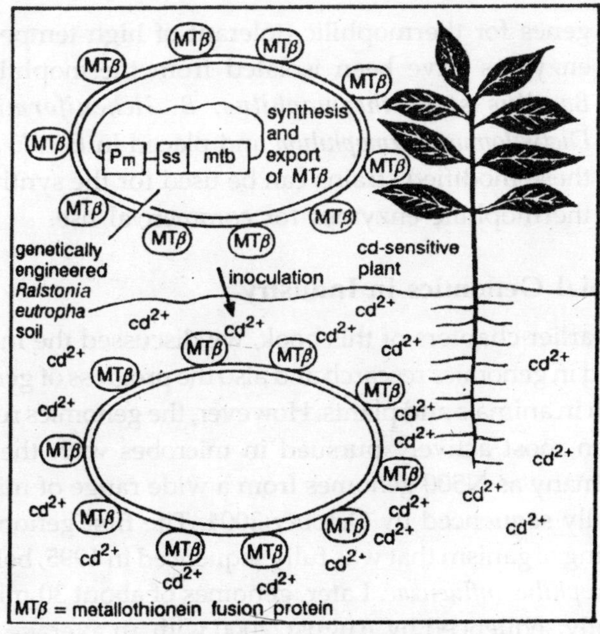

Production of other chemicals and fuels

Recombinant DNA technology is also having a significant impact on microbial production of chemicals and fuels. Following are some of the examples of this utility :

(i) genetically engineered strains of *Bacillus amyloliquefaciens* and *Lactobacillus casei* have been prepared for production of amino acids on a large scale;

(ii) industrially useful bacteria can utilize cheaper feedstocks (substrates) like D-xylulose, lignocellulose or cellulose; for instance, *Zymomonas mobilis* (normally incapable of using lactose) carrying cellulase gene from *Cellulomonas uda* has six fold increase in cellulase activity;

(iii) *E. coli* and *Klebsiella planticola* carrying genes from *Z. mobilis* could utilize glucose and xylose to give maximum yield of ethanol;

(iv) genes for thermophilic (tolerant of high temperature) enzymes have been isolated from thermophiles like *Bacillus stearothermophilus, B. licheniformis* and *Dictyglomus thermophilum* and placed in *E. coli,* so that these modified strains can be used for the synthesis of thermophilic enzymes for commercial use.

Microbial Genomics in Industry

In earlier chapters of this book, we discussed the methods involved in genomics research and also the progress of genomics research in animals and plants. However, the genomics research has been most actively pursued in microbes with the result that as many as N500 genomes from a wide range of microbes were fully sequenced by October 2004. The first genome of a free-living organism that was fully sequenced in 1995, belonged to *Haemophilus influenzae.* Later, genomes of about 30 microbes were fully sequenced by August, 2000 with an average rate of one genome fully sequenced every two months during 1995-2000. Another 50 genomes were sequenced during 2000-2002 giving an average rate of at least two microbial genomes fully sequenced every month during the period. This rate later increased in later due to the availability of high throughput sequencing at a reduced cost.

The microbes that have been taken up for whole genome sequencing include not only the model organism like *Escsherichia coli. B. subtilis, Sacchoromyces cerevisiae* (budding yeast), but also a number of microbes, which either cause dreadly diseases or those, which are used in microbial industry. The latest in this series was a report (in October, 2002) of the whole genome sequence of malarial parasite, *Plasmodium falciparum* along with that to the mosquito *(Anopheles gambiae)*, which is the vector for this parasite, thus spreading the disease (malaria).

Analysis of Microbial Genomes

A comparison of microbial genome sequences has revealed that the gene density in microbes is about one gene per kilobase and that 50% of the genes have unknown functions. It is possible that when the functions of all these genes become known, we may identify entirely novel biochemical pathways, which may be relevant to biotechnology and medicine. It has also been found that 25% of the genes in each species examined are unique, encoding proteins with no sequence similarity with any of the known protein sequences, suggesting the availability of high level of genetic diversity in microbial world. This diversity will be exploited in future for a variety of biotechnological applications.

Analysis of microbial genomes also allowed the study of evolution of nuclear and organellar genomes of higher plants and that of the microbial genomes themselves. For instance, evolution of mitochondrial genome from proteobacteria like *Rickettsia* and that of chloroplast genome from cyanobacteria like *Synechocystis*, has been suggested through the study of genome sequences. An origin of eukaryotic nuclear genome from the fusion of achaebacterial and eubacterial genomes has also been suggested. Another phenomenon discovered through genome analysis is the extensive horizontal or lateral gene transfer, which involves transfer of genes from microbes to the nuclear and organellar genomes of eukaryotes, from organellar genome to the nuclear genome and also the transfer of genes among microbial genomes themselves. A more detailed discussion of this subject is beyond the scope of this book.

Use of Genome Analysis for Designing Vaccines and Drugs

The microbial genome analysis (particularly the study of the genomes of pathogens) will also be utilized extensively for designing diagnostic kits, vaccines and drugs against the diseases caused by these pathogens. This use of microbial genomics has achieved special significance, since it is now known that the pathogens often have the ability to alter their antigenic potential, which enable them to evade the human immune system. This has made it more difficult to design vaccines and drugs by conventional methods, thus making the genome analysis more rewarding. For instance, new vaccine candidate against subgroup B *Neisseria meningitidis* (Men B), on the basis of genome analysis, were reported in the year 2000. Similarly, on the basis of genome sequence of *Plasmodium falciparum* (which causes malaria), two genes were identified, which encode enzymes of DOXP pathway (DOXP = 1-deoxy-D-xylulose-5-phosphate) involved in the synthesis of isoprenoids such as cholesterol. This allowed the designing of inhibitors of these enzymes, which in their turn inhibited the entire DOXP pathway, resulting in the inhibition of the growth of the parasite *P. falciparum*. This approach for development of antimalarial drugs will be further facilitated by the whole genome sequences of *P. falciparum* and *A. gambiae*, which became available in October, 2002.

Genomics of Agricultural Microbes (plant pathogens and symbionts)

Although the first microbial genome was sequenced in 1995, the first agricultural microbe, a citrus pathogen *(Xylella fastidiosa)* was sequenced only in the year 2000. During 2000-2002, several agricultural microbes were used for whole genome sequencing. Some of these microbes include the following :

(i) *Ralstonia solanacearum* (a chromosome of 3.6 Mb + a plasmid of 2.1 Mb) that causes bacterial wilt disease in more than 200 plant species belonging to 40 different families;

(ii) two species of *Xanthomonas (X. axonon podis* pv *citri,* formerly known as X. *campestris,* a well known pathogen of crucifers), infecting a variety of crop plants including rice, cotton, citrus, grape and bean;

(iii) *Magnaporthe grisea,* an ascomycetes, causing rice blast,

(iv) *Phytopthora* spp *(P. infestans* causing late blight of potato and *P. sojae,* causing root and stem rot of soybean), and

(v) *Rhizobium melilotii,* a well known symbiotic bacterium. The information generated from whole genome sequencing of these microbes (including fungi) will prove useful not only in developing strategies for plant protection and disease management, but also in designing strategies for beneficial plant-microbe interactions for improved productivity.

9

Production of Haploids

Haploids are defined as sporophytes with gametophytic chromosome number and have been produced in a variety of plant species using a variety of methods. Although, the significance of haploids in genetics and plant breeding has been recognized for long time, it has received renewed emphasis with the advent of new biotechnology, so that the production of haploids has become an important component of biotechnology programmes in different countries. Although, haploids could be produced following delayed pollination, irradiation of pollen, temperature shocks, colchicine treatment and distant hybridization, the most important methods currently being utilized under biotechnology programmes include :

(i) anther or pollen culture and ovule culture and

(ii) chromosome elimination following inter-specific hybridization (bulbosum technique). These techniques for production of haploids and the uses of haploids in genetics and plant breeding will be discussed in this chapter.

Production of Haploids

Anther and Microspore Culture : One of the very popular methods for production of haploids is through culturing anthers or microspores on artificial culture medium. This leads to the growth of microspores into sporophytes. After the initial reports of successful production of haploids from anther culture in *Datura* (Guha and Maheshwari , 1966, 1967), haploids have been obtained in more than 150 species belonging to more than two dozen families of angiosperms. These include a wide variety of economically important species.

More often, anthers rather than microspores are cultured, since methods for extraction and culture of microspores differ and have been successful only in a few species *(Datura inoxia, Nicotiana sylvestris, N. tabacum, Oryza sativa, etc.).* The following parameters have been recognized as particularly important for successful anther and microspore culture; (i) conditions of growth of donor plant; (ii) genotype of donor plant; (iii) the pretreatment; (iv) the developmental stage of anther/ microspore; (v) the culture medium and the conditions during culture growth.

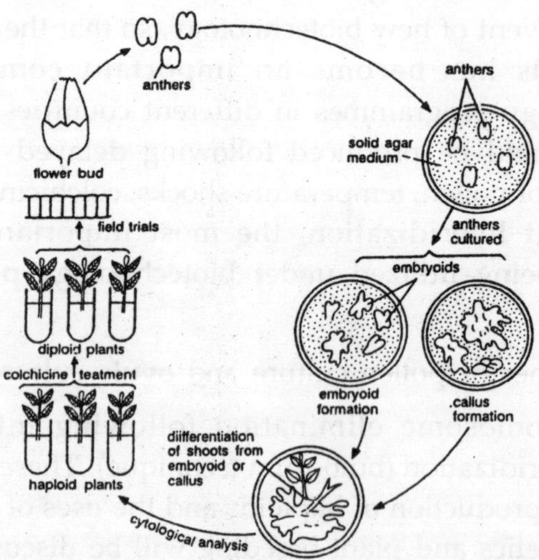

Different steps involed in the production of haploid plants using anther culture.

The different steps involved in anther and microspore cultures are shown in Figures. In either case, flower buds are brought to a laminar flow chamber and sterilized using appropriate chemical treatment before their dissection. While removing anthers from the flower buds, care is exercised to avoid injury, because injury leads to development of callus, giving a mixture of diploids, haploids and aneuploids. As shown in Figure, the anthers are generally cultured on solid agar medium, where they may directly give rise to embryoids or may lead to callus formation, before differentiation. The embryoids develop into haploid plantlets, which are colchicine treated to get doubled haploid (DH) homozygous plants to be field tested for selection.

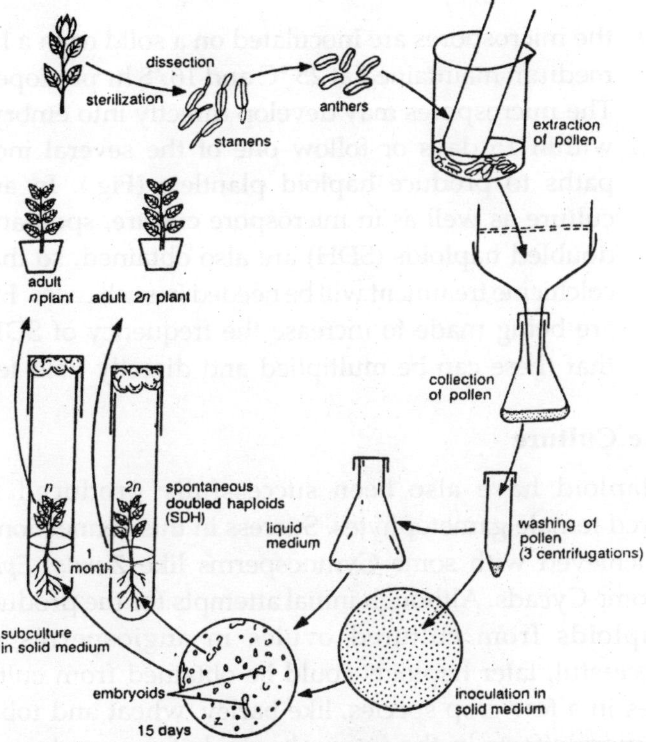

Different steps involved in the production of haploid plants using isolated pollen grains.

Microspore culture may be preferred over anther culture, even though the degree of success is low in this case. Microspores are collected first using the following steps :

(i) anthers, dissected as above, are taken in a sterile beaker containing liquid medium, and are pressed with a glass rod or syringe piston to allow the microspores squeeze out;

(ii) the suspension with anthers and microspores is filtered through a nylon sieve, which allows only the microspores to pass through;

(iii) the filtrate is centrifuged thrice, for 5 minutes each, at 500-800 rpm (the pellet resuspended in a fresh medium each time);

(iv) the microspores are inoculated on a solid or in a liquid medium maintained at 25 °C and 16/8 hr photoperiod. The microspores may develop directly into embryoids within 15 days or follow one of the several indirect paths to produce haploid plantlets (Fig.). In anther culture as well as in microspore culture, spontaneous doubled haploids (SDH) are also obtained, so that no colchicine treatment will be needed in such cases. Efforts are being made to increase the frequency of SDH, so that these can be multiplied and directly field tested.

Ovule Culture

Haploid have also been successfully produced from cultured female gametophytes. Success in this connection was first achieved with some Gymnosperms like *Zamia, Ephedra* and some Cycads. Although initial attempts for the production of haploids from cultured ovules in angiosperms were unsuccessful, later haploids could be obtained from cultured ovaries in a few crop species, like barley, wheat and tobacco. With more efforts in the future, haploids from ovule culture will be produced in a number of plant species. Further research is, therefore, necessary to improve our understanding about the following: (i) the events leading to induction of haploidy in female gametophytes, (ii) the factors which control the *in vitro* development of proembryo into the fully organized plants and (iii) the differences in the growth patterns of *in vitro* development of unfertilized female gametophytes (in ovules) and the male gametophytes (in pollen).

Chromosome Elimination

Somatic Reduction and Chromosome Elimination : Cases are known where either spontaneously or due to specific treatments, the chromosome number was reduced to half in the somatic tissues, a phenomenon described as somatic reduction or reductional mitosis.

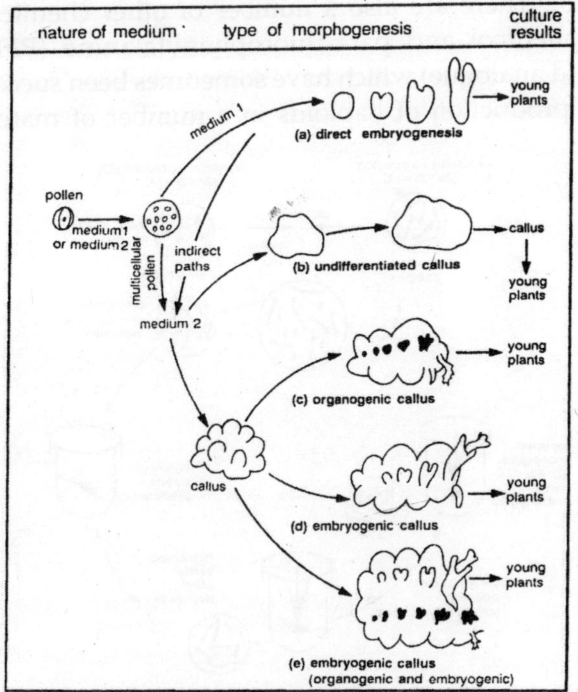

A direct and several indirect paths leading to production of haploid plants from isolated pollen grains.

Swaminathan and Singh (1958) induced a haploid branch on a watermelon by irradiation of the seed used. This must have resulted by the reduction of chromosome number in the somatic tissue through an unknown mechanism (perhaps due to spindle organizer abnormalities). Similarly, colchicine treatment has also been found to lead to the reduction in chromosome number. Franzke and his associates at South Dakota State University (USA) conducted a series of interesting studies in *Sorghum vulgare.* They found that in *Sorghum vulgare,* somatic tetraploid *(2n = 4x)* cells responded to colchicine treatment and gave rise to diploid cells, which took over the growing point completely thus giving rise to diploid (dihaploid) individuals. In general, attempts to induce somatic reduction in other species by treating seeds or seedlings with colchicine have failed or met with only partial success as in flax (Dirks *el al,* 1956). There are also a number of other chemicals like chloramphenicol and para-fluorophenylalanine (PFPA, an amino acid analogue) which have sometimes been successfully used for production of haploids in a number of materials.

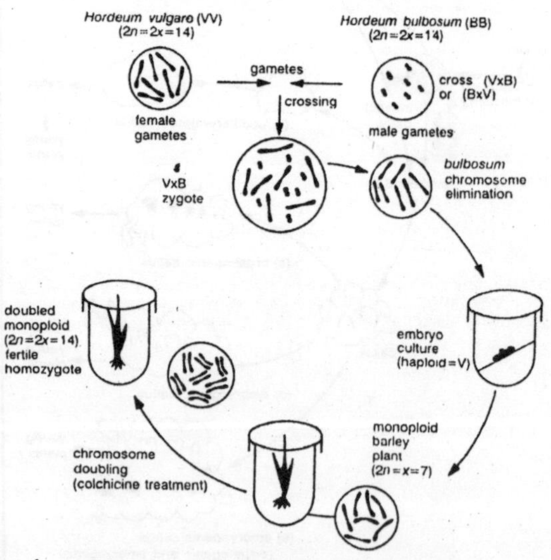

Scheme for production of monoploid and doubled monoploid barley plants using wide hybridization (bulbosum method involving elimination of Hordeum bulbosum chromosomes).

Barley haploids through crosses with *Hordeum bulbosum*. A situation analogous to somatic reduction, occurs when in an interspecific or intergeneric hybrid, chromosomes of one parental genome are preferentially eliminated. This method has been extensively utilized for production of haploids in barley and also in wheat. The production of haploids in barley in this manner involves inter-specific hybridization of *Hordeum vulgare* (2n =14) with *H. bulbosum* (2n=14), a cross pollinated, self incompatible perennial species (Fig.) Most progeny (95%) are barley haploids, remainder being diploid hybrids. A cross between 2*x* barley and 4x *H. bulbosum* mainly produces triploid hybrids. Cytological examination of early embryo and endosperm in VV × BB or BB × VV hybrids revealed gradual elimination of chromosomes (Table). While most genotypes of barley readily produce haploids, chromosome elimination seems to be controlled by genotypes of both parents.

Results of Crosses Between Hordeum Vulgare (2x = VV, 4x = VVVV) and H. Bulbosum (2x = BB,4x = BBBB) (After Lange, 1971).

Cross	*Progeny*		
	Hybrid	:	*Haploid ratio*
VV × BB	2 (VB)	:	1 (V)
BB × VV	4 (BV)	:	2 (V)
VV × BBBB	145 (VBB)	:	1 (V)
BBBB × VV	12 (BBV)	:	0 (V)
VVVV × BBBB	11 (VVBB)	:	15 (VV)
BBBB × VVVV	5 (BBVV)	:	3 (VV)

Kasha and Kao (1970) summarized the findings for haploid production using the above technique as follows :

(i) a very high frequency of haploids can be obtained and these haploids are not associated with any diploid hybrids;

(ii) these haploids can be obtained with any cultivated barley and would resemble cultivated barley genome in reciprocal crosses.

Contrary to these observations, Lange (1971) showed that haploids obtained due to chromosome elimination are associated with' hybrids (Table). With embryo culture technique, he was able to obtain viable hybrids from several cross combinations, although frequency of hybrids was low except in case of VBB hybrids, which were quite frequent.

Following is the subsequent summary of characteristics associated with chromosome elimination, keeping in view the above results (Kasha, 1974).

(i) Elimination occurs in hybrid embryos following interspecific or intergeneric hybridization.

(ii) Both haploids and species hybrids may be produced from a cross, and the haploids can have female or male origin, so that androgenetic haploids can also be produced.

(iii) Genotypic variation may be found within parental species and different species may differ in their response to chromosome elimination.

(iv) Initial seed set is good followed by shrivelling, which requires embryo rescue.

(v) Hybrids obtained in such crosses exhibit chromosome instability and aneuploidy in somatic and meiotic cells.

Haploids of other *Hordeum* species from interspecific hybrids Chromosome elimination is also quite common among other *Hordeum* inter-specific hybrids not involving *H. vulgare,* and has been reported in about 20 such hybrids. For instance, *H. bulbosum (4x)* × *H. secalinum (Ax)* gave haploids of *H. bulbosum,* while *H. secalinum* (4x) x *H. bulbosum* (2 × and 4x) and *H. secalinum* (4x) × *H. vulgare* (2x) gave haploids of *H. secalinum* along with hybrids.

Wheat Haploids through crosses with *H. bulbosum.* Haploids in hexaploid wheat (var. Chinese Spring) can also be produced through chromosome elimination in hybrids of wheat with *H. bulbosum* (both *2x* and *4x). A* frequency of 13.7% grain set in hybrids with *2x bulbosum* and 43.7% grain set with *4x bulbosum*, were obtained. In order to explore the possibility of further use of this technique in wheat breeding programmes, studies were conducted at Cambridge (U.K.) on the crossability of wheat with *H. bulbosum.* It was found that most of the European varieties did not cross, but four of the six Australian varieties were successfully crossed.

A similar crossability pattern of wheat varieties was observed with rye also so that it was concluded that crossability with rye and *H. bulbosum* was controlled by same genetic factors. Through the use of chromosome substitution lines, it was also shown that in the variety Hope 1B, 1D, 2D, 3A, 3D, 4A, 4B and 6D have genes promoting crossability, while 5A and 5B have genes reducing crossability. In view of the limitation of the crossability of *H. bulbosum* with only few wheat varieties, *H. bulbosum* can not be extensively used for production of haploids in common wheat.

Wheat Haploids from Wheat × Maize Crosses : M.D. Bennett and his coworkers in Cambridge, U.K. (now at Kew, U.K.) demonstrated that wheat × maize crosses can be made successfully, and hybrid embryos can be rescued. Due to elimination of maize chromosomes, wheat haploids can be obtained. Wheat × maize crosses are now extensively used at CIMMYT, Mexico and elsewhere for the production of haploids in bread wheat, particularly with an objective of producing doubled haploids (DH) for developing mapping populations.

Mechanism of Chromosome Elimination : The Mechanism of chromosome elimination in *Hordeum* hybrids was studied in some detail and following conclusions were made:

(i) normal double fertilization occurs in inter-specific crosses as confirmed through cytological study;

(ii) after fertilization, there is a gradual and selective elimination of *H. bulbosum* chromosomes from nuclei of endosperm as well as embryo cells, so that eventually haploid embryos are produced. No evidence of the elimination of *H. vulgare* chromosomes was available.

Differences were, however, observed in the rates of chromosome elimination and these were attributed to genotypic differences among parents used in hybridization programmes. The possible causes of chromosome elimination included :

(i) sudden short-age of proteins in developing embryo and endosperm, and

(ii) the better ability of *vulgare* chromosomes, relative to *bulbosum* chromosomes, to form spindle attachments. Other possible causes like differences in mitotic cycle, congression during mitosis, etc. were ruled out. It has also been demonstrated that the elimination of *bulbosum* chromosomes is under genetic control. The use of primary trisomics and monotelotirsomics in crosses with tetraploid *H. bulbosum* suggested that both arms of chromosome 2 and the short arm of chromosome 3 of *H. vulgare* are responsible for chromosome elimination although their effect may be neutralized or offset if sufficient dose of *bulbosum* chromosomes is available. The effect of these *vulgare* chromosomes, however, will be indirect.

Detection or Identification of Haploids

Morphology : A number of methods have been developed for the detection of haploids in a population of diploids. Although the most effective method is the use of genetic markers to be discussed in the next section, phenotypic effects can also be used for identification of haploids. The size of vegetative and floral parts and also that of the cells in haploids are reduced relative to diploids. In maize some success was achieved in the identification of haploids on the basis of coleoptile thickness. Stomata size, pollen size and pollen abortion can be other criteria to help initial screening, although in each case cytological confirmation will be essential.

Genetic Markers : Many genetic marker systems are used for recognizing the haploids. The basic principle involved is that the hybrids and nonhybrids should differ, so that any failure of fertilization and parthenogenetic development of embryo can be recognized in the progeny at the seed stage, seedling stage or adult plant stage. For instance, if a female parent in a cross carries a recessive marker and is pollinated by dominant stock, then any recessive progeny would be generally a maternal haploid, although possibility also exists of its being a diploid of one of the following types:

(i) diploid hybrids, with mutation or deficiency for the marker gene in male parent,

(ii) diploid hybrids with suppression of dominant marker, due to a suppressor gene or due to a disease,

(iii) diploids which are maternal due to apomictic development.

If the identification of haploids needs to be done at the seed stage, then the marker should express itself in the late stage of embryo development, such as aleurone colour marker. If screening is to be done at seedling stage, then the marker should clearly express in the seedling (e.g. the purple seedling colour). This technique of genetic marker has been extensively used in developing maize monoploids. Most extensively used markers included 'a_1' for brown coloured aleurone, plant and root and *lg* for liguleless, their corresponding dominant markers being *'A'* for purple colour and *'Lg'* for liguled character. If the female parent carries *'a'* and/or *lg*, then it crosses with male parent carrying *'A'* and/or *'Lg'*, the monoploids will be recognized by lack of purple colour and/or ligule. Another critical marker gene is 'R' allele *($R^{nj\text{-}cudu}$)*, which produces deep pigmentation of aleurone, and thus allows screening at the dormant seed stage. Other markers included purple plumule *(A, Pu_1 Pu_2)*, scutellum colour *(A, C, R* genes, and S_1, S_2 and S_3 genes), and purple coleorrhiza *(Pc_1Pc_2,Pc_3* and *Pc_4)*. These markers did not prove to be equally useful, because the frequency of these markers in maize populations is otherwise also quite high. Endosperm markers, however, are not as useful because very often monoploid embryo in the seed is associated with triploid endosperm, so that the latter will exhibit the male dominant endosperm trait. It should also be noted that for the recognition of paternal or androgenetic haploids, the dominant marker should be present in the female parent so that the androgenetic haploids may be phenotypically different from the hybrid.

In most of the above cases when genetic markers are used, 90-98% of the progeny can be discarded rather early and a very high frequency of haploids are recovered in the remaining population. The above technique of using genetic markers is particularly useful for the development of inbred lines in maize, but in crops like barley and wheat which are self pollinated and where doubled haploids need to be used for cultivation, the availability and utility of genetic markers may be limited.

Use of Haploids in Plant Breeding Programmes : Among the different uses of haploids, their use in crop improvement is considered to be the most significant and stands out as the most important reason for emphasis on haploid research. Some of the important crops, where haploid plants have been produced are listed in Table.

Some Commercially Important Crops where Haploids have been Produced.

1.	Alalfa	9.	Cucumber	17.	Rye
2.	Asparagus	10.	Grape	18.	Strawberry
3.	Barley	11.	Kale	19.	Sugarcane
4.	Broccoli	12.	Peanut	20.	Sunflower
5.	Citrus	13.	Pepper	21.	Sweet potato
6.	Clover	14.	Potato	22.	Tomato
7.	Coffee	15.	Rapeseed	23.	Triticale
8.	Com	16.	Rice	24.	Wheat

The methods used for utilizing haploids in plant breeding depend on the inheritance pattern observed in a crop, particularly when we use haploids derived from a polypldid crop. For instance allopolyploids, particularly those which are self pollinated (e.g. tabacco), exhibit disomic inheritance, while autopolyploids (e.g. potato), particularly those, which are cross pollinated, exhibit polysomic inheritance. Uses of haploids for two types of crops will be discussed separately.

Use of Haploids in Species with Disomic Inheritance : Production of homozygous lines. The most important use of haploids, often advocated, is their use in the production of homozygous lines, which may be directly used as cultivars or may be used in breeding programmes. For instance, a doubled haploid, 'Marglobe tomato' has been used commercially. Similarly, doubled haploids were used as inbred lines in hybrid maize programme. However, the chief limitation in the use of haploids for production of homozygous lines had been the low frequency, with which they were available. In recent years, the availability of anther culture technique and the use of crosses with *Hordeum bulbosum* in barley, and with maize in wheat have led to the production of haploids in high frequency. This has greatly accelerated the pace of research for production of haploids and their use for getting doubled haploids (DHs).

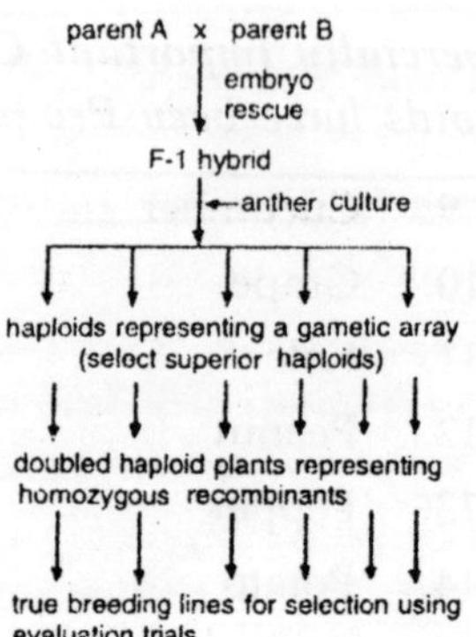

Different steps in hybrid sorting using anther culture technique.

Hybrid Sorting in Haploid Breeding : One of the essential steps in haploid breeding involves selection of superior plants among haploids derived from F_1 hybrids through anther culture. It is popularly described as hybrid sorting and virtually means selection of recombinant superior gametes (Fig.).

The haploid method of breeding involving hybrid sorting is considered superior over pedigree and bulk methods, *firstly* because the frequency of superior gametes is higher than the frequency of corresponding superior plants in F_2 generation, and *secondly* because haploid breeding reduces significantly the time required for development of a new variety (see later). For instance, if we assume that the frequency of superior F_1 gametes is one in one hundred (1/100), then the frequency of homozygous F_2 plants derived from the fusion of two such superior similar gametes would be one in ten thousand (1/10000). Therefore, smaller populations of doubled haploids derived from haploids will need to be handled.

The above technique of hybrid sorting has been successfully utilized, particularly in China and Japan, for development of new varieties in several crops (Table). In Japan, an excellent tobacco cultivar, 'F-211' bred through anther culture, is resistant to bacterial wilt and has a mild smoking quality. In China, more than 100 rice varieties developed using the same technique, gave an increased yield and, therefore, occupy significant areas of rice cultivation (Table).

Some High Yielding Crop Varieties Through Hybrid Sorting Using Anther Culture.

Crop	Varieties developed	Attributes
1. Tobacco *(Nicotiana tabaccum)*	F 211, Tanyu 1, Tanyu 2, Tanyu 3	Disease resistant, mild smoking
2. Rice *(Oryza sativa)*	Tanfong 1, Tanfeng 1, Tanyu 1, Huayu 1, Huayu 2, Huayu 15, Hua-03, Hua Be 702 Hua Han Zao, Zhonghua 8, Zhonghua 9, Zhonghua 10, Zhonghua 11, Tonghua 1, Tonghua 2, Nanhua 11, Chaohua Ai, Ta Be 78, Shin Shu, Xin Xiu, Xin Xion, Late Keng 76, Late Keng 959, Zhe Keng 79-66, Huapei Shanyou 63	Good quality, disease resistant, high yield
3. Wheat *(Triticum aestivum)*	Huapei 1, Lunghua 1, Jinghua 1, Yunhua 1, Yunhua 2	Rust resistant, cold resistant, high yield, large spikes, more tillers, resistant to rust & powdery mildew, lodging resistant
	Jingdan 2288 (winter wheat)	Large spikes, more tillers, resistant to rust & powdery mildew, lodging resistant

Some High Yielding Crop Varieties Through Hybrid Sorting Using Anther Culture

Variety	Cultivated acreage (ha)	Yield (t/ha) Cal	Increase in yield over local variety (%)
1. Xin Xion	$>3\times10^4$	—	—
2. Huayu 1	1×10^4	—	—
3. Huaya 15	—	8—11	—
4. Late Keng 76	1×10^4	6.7	17
5. Late Keng 959	$>3\times10^4$	4.5–6.0	10
6. Tonghua no.1 and 2	2×10^4	6.8–7.7	10–15
7. Hua Han Zao	1.5×10^5	—	8.7–27.7
8. Zhe Keng 79-66	1×10^5	6.05	8.1
9. Zhonghua 8, 9, 10, 11	1.5×10^5	7–10	8.8–13.7
10. Nanhua 11	1×10^4	6.12	7.4–10.1
11. Chao Hua Ai	—	6.5–8.3	18.8–51.7
12. Huapei Shanyou 63	3×10^5	—	5.7
13. Hua-03	—	6.0	10 (high protein)

Haploids have also been successfully utilized for breeding of barley, maize, sugarcane, oilseed rape or canola *(B. napus)*, bell peppers and some other crops. In barley, atleast two varieties (Mingo, Gwylan) have been produced in Canada using haploids derived through *bulbosum* technique of chromosome elimination.

A Comparison of Pedigree and Bulk Methods of Conventional Plant Breeding with Haploid Breeding in a Self Pollinated Crops Like barley.

Years	Pedigree method	Bulk method	Haploid method
1.	Cross parents	Same as in pedigree method	Cross parents
2.	Grow Fl		Grow Fl; derive haploids and DHs (anther culture)
3.	Grow F2 (2000-10000 seeds space planted); select for qualitative traits; harvest seed on single plant basis		doubled haploids (DHs) are used for initial trials
4-5.	Grow F3, F4 (plant to progeny rows from selected plants); select rows, plants	Harvest seed in bulk in F2-F5; no selection; only natural selection	Trials to verify performance; release of variety
6.	Grow F5; initial trial (3-4 rows of each plant progeny); compare and select.		
7-9.	F6-F8; trials to verify performance	Space planting; select single plants; plant to progenies; yield trials	
10.	Final trial and release of variety	Final trial and release of variety	

The varieties were superior for characters like disease resistance (against powdery mildew and barley yellow mosaic virus) and other yield contributing traits. Similarly, in maize, superior inbred lines were developed through screening of haploids derived from anther culture in a hybrid. In sugarcane, selection among anther-culture derived haploids led to the development of superior lines with tall stem and higher sugar content. In *Brassica napus* also, anther derived doubled haploids, produced in Canada, had the desirable low erucic acid and glucosinolate content. In another case, a haploid derived from twin seedlings in *B. napus* led to the development of a superior cultivar 'Maris Haploma'. In bell peppers, DH lines produced from anther-derived haploids, exhibited all shades of colour ranging from dark green found in one parent to light green found in the other parent of the hybrid employed.

Despite the above conspicuous achievements made through haploid breeding, we find that on a world wide level, the acreage occupied by varieties developed through haploid breeding is negligible. In India, no variety in any crop has so far been developed through haploid breeding. Even at an international level the emphasis on haploid breeding has gone down in recent years, partly due to failure in achieving the expected targets and partly due to shift in emphasis toward molecular plant breeding involving transgenic crops and molecular marker aided selection.

A comparison of haploid breeding with conventional breeding. A comparison of haploid method of breeding with conventional methods (pedigree method, bulk method, single seed discent or SSD) suggests that the time period required in haploid breeding is only 4-5 years, as against 10 years required in pedigree/bulk method. Production of spontaneously doubled haploids for breeding. The utility of haploids in plant breeding eventually requires production of doubled haploids (DH) as discussed above. For this purpose, often efficient colchicine treatments are applied and colchicine doubled haploids (CDH) are obtained. In a system where haploids are detected through genetic markers, we can also identify the spontaneously doubled haploids (SDH), if doubling of chromosomes does occur spontaneously. This will avoid unnecessary colchicine treatment and the breeder can save time and plant material.

Although there are several reports of SDHs produced during haploid production, but at Svalov in Sweden, *hap* gene was utilized, not only for the production of haploids, but also for the production of SDHs (Hagberg and Hagberg, 1987). It utilized a marker gene system for mildew resistance due to an allele *(ml-o)*, which is recessive to susceptibility allele *(Ml-o)*.

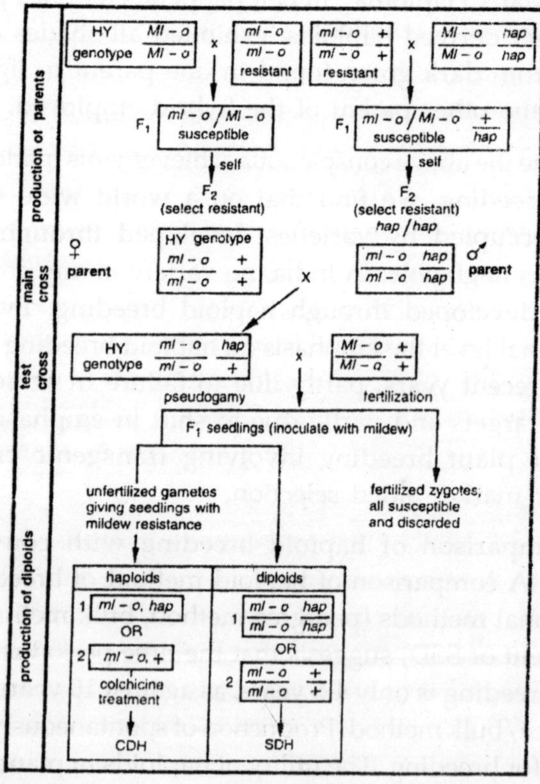

The Use of hap Gene in Barley for the Production of Colchicine Doubled Haploids (CDH) and Spontaneously Doubled Haploids (SDH) as Used at Svalov, Sweden.

The breeding scheme used to produce spontaneously doubled haploids is shown in Figure. In such a programme, spontaneously doubled haploids (SDH) were three times as many as the normal haploids, which seems to be really promising. These SDH plants were found to be completely homozygous thus proving their maternal origin due to doubling of chromosome number in the original egg cell. The marker *ml-o* is suitable, because it allows screening at the seedling stage, but other gene pairs like *K-k* (hooded *vs* normal awns) or V-v (two row *vs* six row ears) can also be used in barley. It is expected that more work in this direction will be carried out in future to make SDH production system useful to the breeders and cytogeneticists.

Use of Haploids in Species with Polysomic Inheritance : Polyhaploids derived from polyploids exhibiting polysomic inheritance may be homozygous or heterozygous. For instance, dihaploids of potato may be *AA* or *Aa* or *aa*, depending upon the genotype of the tetraploid potato, which can be heterozygous for this gene *(AAAa, AAaa, Aaaa)*. Therefore, by doubling of haploids, homozygosity can not always be achieved in these cases, unless dihaploids are again used for production of monoploids. Therefore, the use of haploids in an autotetraploid crop like potato differs from that in crops like barley, rice, wheat, etc.

Analytical Breeding in Potato : In a crop like potato, the genetic improvement through crosses and selection may be achieved at the dihaploid level and improved dihaploids may then be used for chromosome doubling to get increased tuber size. This has been described as analytical breeding (Chase, 1963), which has the following advantages:

(i) disomic inheritance in the dihaploids provides simpler inheritance ratios, ease of selection and smaller population size required to recover the recombinants;

(ii) dihaploids can be crossed with wild diploid species for introducting new genes;

(iii) the dihaploids, represent a level of homozygosity equivalent to three generations of self pollination in an autotetraploid.

Two Breeding Schemes Using 2n Gametes in Potato : In the above analytical method, where, improved dihaploids are doubled by colchicine, there is reduced vigour of dihaploids due to inbreeding and expression of deleterious genes. This leads to failure in recovery of sufficient vigour in the tetraploids on chromosome doubling, because vigour depended on intra and interlocus interactions in multiple allelic condition in the autotetraploids. To overcome this problem, following two breeding schemes were suggested by Mendiburu and Peloquin (1977) :

(i) $4x \times 2x$, where $4x$ is a cultivar and $2x$ is a hybrid between dihaploid and a wild diploid species. The cross gives $4x$ progeny due to functioning of $2n$ pollen in the $2x$ parent.

(ii) $2x \times 2x$, where both parents are hybrids derived from crosses between dihaploids and wild species.

The $4x$ progeny in this case is derived due to functioning of $2n$ eggs in one parent and 2n pollen in the other parent. Since, in both these schemes, $2x$ parent is not a dihaploids but a hybrid (haploid × species), it is apparent that the dihaploids are first used for enriching the germplasm at the $2x$ level and then the enriched $2x$ is used for its transfer to $4x$ ploidy level. Due to disomic inheritance, there are several advantages in using $2x$ hybrids derived from '$2x$ haploid Tuberosum × $2x$ wild species'. In the first of the two schemes discussed above, i.e., in $4x \times 2x$ crosses, $2x$ hybrid rather than $2x$ wild species is used directly with $4x$ potato cultivars. Since the 4* cultivar is an adapted genotype, a cross with $2x$ wild species will lead to a loss of 50% of the adapted germplasm, which will be replaced by unreduced *2n* gamete from *2x* species. On the other hand the *2x* hybrids as above have only 50% of germplasm from *2x* wild species, so that the *4x* progeny (derived from *4x* cultivar × *2x* hybrid) will have only 25% unadapted germplasm from the wild species.

Analytical Synthetic Breeding for Potato : Another approach described as 'analytical synthetic breeding scheme' was proposed by Wenzel *et al.* (1979). This takes advantage of protoplast fusion and plant regeneration to bring together different genotypes into a completely heterozygous tetraploid potato. In this approach, dihaploids are used again for the production of monoploids, before the tetraploid is resynthesized. The breeder can evaluate and select the genomes present in monoploids. The selected monoploids are then used for resynthesis of tetraploids through protoplast fusion and regeneration as follows: (monoploid$_1$ + monoploid$_2$) + (monoploid$_3$ + monoploid$_4$). The selection of best somatic hybrids may be feasible on the basis of their growth rate in cell cultures.

The above scheme of analytical synthesis can be easily used in potato, because once the resynthesized tetraploids are regenerated, they can be multiplied via tubers for commercial cultivation. In other autotetraploid crops like alfalfa *(Medicago sativa)* or orchard grass *(Dactylis glomerata)*, where reproduction is by seed, the above scheme will not be successful. However, the, analytical breeding, described earlier can be used. In alfalfa, where *2n* gametes function, the two schemes suggested by Mendiburu and Peloquin (1977) for potato can also be used.

Limitations to Haploid Breeding

As discussed above, haploid breeding has not given the desired and expected dividends, despite substantial investments made on haploid research during 1970-2000. Following are some widely recognized limitations :

(i) Haploids can not be obtained in high frequency required for selection.

(ii) The cost-benefit ratio in haploid breeding is often not favourable, thus discouraging the use of haploid breeding despite its obvious advantages.

(iii) Haploids will express recessive deleterious traits, and additional deleterious mutations may arise during anther culture.

(iv) Different ploidy levels may be available so that haploid status may need to be confirmed cytologically; alternatively pollen culture may need to be used, which is more expensive and gives relatively low success rate.

Other uses of Haploids

Mutation Research : The teachnique of cell culture (including haploid production) is also used for induction and selection of mutations. The use of haploids provide with the following additional advantages in mutation research :

(i) Due to the presence of only one set of chromosomes, even the recessive mutation are immediately expressed in haploids (unlike diploids, where the recessive mutations express only in segregating single plant progenies grown in M_2 generation).

(ii) Pollens can be plated on solid medium like single cells and therefore much bigger populations can be screened in the laboratory, without initially growing the plants in the field.

In several crops, desirable mutants have been isolated among haploids derived in culture. These desirable attributes included resistance to drugs (e.g. antibiotics), amino acid analogues, pathotoxins, salts, herbicides, chilling, diseases (caused by fungi, bacteria, viruses and nematodes). Such mutants can be used for a variety of purposes leading to the development of suitable cultivars.

Gametoclonal Variation

Somaclonal variation, which originates in tissue culture experiments and can be profitably utilized for improvement of economic traits in crop plants (For more details see author's another text book "Biotechnology & Genomics"). Similar variation can also be observed among haploid plants (having gametic chromosome number), developing from cultures derived from anthers or microspores. Since this variation is available among plants with gametic chromosome number, it is popularly described as gametoclonal variation. The utilization of gametoclonal variation differs from that of somaclonal variation in the following respects :

(i) recessive mutation will be immediately expressed while somaclones need to be selfed in some cases to allow expression of variation;

(ii) chromosome number needs to be doubled in selected gametoclones. The gametoclonal variation is actually utilized in hybrid sorting during the use of haploids for plant breeding purposes.

Cytogenetics Research

Haploids have been found to be useful in various areas of cytogenetics research, including the following :

(i) production of aneuploids,

(ii) determination of basic chromosome number and

(iii) determination of the nature of polyploidy.

For instance, in wheat initially monosomics were produced by E.R. Sears by pollinating a haploid wheat plant by the pollen from a normal wheat plant. The monosomics in this case presumably resulted due to functioning of eggs with restitution nuclei deficient for any individual chromosome. In potato, haploids were used for the production of trisomics using one of the following procedures :

(i) isolation of parthenogenetic aneuhaploid offspring following $4x \times 2x$ crosses;

(ii) isolation of aneuploids from $3x \times 2x$ crosses, where $2x$ is a haploid. In both the schemes, a number of trisomics with 2n = 25 could be obtained.

In tobacco, several nullisomics were derived from haploids obtained from monosomics, which could not produce nullisomics on selfing due to sterility. Haploids have also been used for the production of alien barley addition and substitution lines in a crop like wheat.

A study of meiosis in haploids also gives evidence for the original basic chromosome number in a species or a genus. For instance, in pearl millet *(Pennisetum typhoides),* occurrence of some pairing (upto two bivalents in some cells) suggested that the basic chromosome number may be $x = 5$. Similar evidence is also available in maize, which is, therefore, considered to be an archaic tetraploid. In potato also, chromosome behaviour of dihaploids (2n = $2x$ = 24) suggested that the base number is x=12 and that potato is an autotetraploid. Chromosome behaviour of monoploids in potato (2n = x = 12) formed only univalents confirming this conclusion.

Evolutionary Studies

A comparison of dihaploids in potato with diploid wild species proved useful in deriving conclusions about the origin of cultivated potato. Since, production of $2n$ gametes is common both in dihaploids as well as in diploid species, it is suggested that Andigena, the progenitor of Tuberosum must have originated many times from different but related $2x$ populations in different locations through sexual polyploidization. Haploids in potato have also been used for crosses with diploid species to conduct genome analysis. The meiosis in F_1 hybrids and further generations suggested that wild species differ from cultivated potato for a small number of genes.

Potato dihaploids were also used for a study of relationship between potato and tomato. A study of pachytene chromosomes in potato dihaploids and tomato (diplods, $2n = 24$) suggested close evolutionary relationship between potato and tomato. Recently potato dihaploids were used for preparing RFLP (restriction fragment length polymorphisms) maps in potato and for a comparison of RFLP maps in potato and tomato. The results indicated a high degree of conservation of DNA sequences.

Genetic Studies

Disomic inheritance and reduced chromosome number in polyhaploids (e.g. dihaploids in potato) offer special advantage for their use in a variety of genetic studies. Using potato as an example, the uses of haploids in genetic studies can be listed as follows :

(i) Using haploids for a study of isozymes through electrophoresis, a number of genes for several isozymes were identified in potato. Progress has also been made in determining the number of genes for other traits like disease resistance and other agronomic traits.

(ii) Using haploids in potato, heritability was also determined for several traits like specific gravity, glucose concentration, tuber dormancy, total yield and tuber weight.

(iii) The genotype of a tetraploid can also be determined by extracting and analysing the possible haploids, as has been done in potato for acid gel proteins, several isozymes and self incompatibility, because the recessive alleles are unmasked in these haploids.

(iv) Intralocus interactions could be studied using haploids in potato, because at a tetra-allelic locus (A_1 A_2 A_3 A_4) in potato, one expects six first order interactions (A_1 A_2, A_1A_3, A_1A_4, A_2A_3, A_2A_4, A_3A_4), three second order interactions ($A_1A_2A_3$, $A_1A_2A_4$, $A_2A_3A_4$) and one third order interaction ($A_1A_2A_3A_4$). As opposed to this, dihaploids will have only one interaction (A_1A_2), which partly explains loss of vigour.

(v) Recently dihaploids × wild species crosses were also used for the preparation of RFLP linkage maps in several crops.

Bibliography

Ahmad, K. Henikoff, S. : *Epigenetic Consequences of Nucleosome Dynamics,* MIT Press, Cambridge, 2002.

Altaian, S. : *Transfer RNA,* MIT Press, Cambridge, 1978.

Anderson, S. : *Sequence and Organization of the Human Mitochondrial Genome Nature,* Oxford University Press, London, 1981.

Andre, C. A. Levy and V. Walbot : *Structure of Plant Mitochondrial Genomes,* John Wiley, London, 1992.

Atkins, J. F. : *Recoding : Dynamic Reprogramming of Translation,* Oxford, New York, 1996.

Bennett, R. H. : *Transcription, Processing and Editing in Plant Mitochondria,* Oxford, New York, 1992.

Bloom, L. B. : *Biochemical Basis of DNA Replication Fidelity,* MacMillan, London, 1993.

Branden, C. and J. Tooze : *Introduction to Protein Structure,* Garland, New York, 1991.

Darnell, J. E., H. F. Lodish and D. Baltimore : *Molecular Cell Biology,* Scientific American Books, New York, 1995.

Das, A. : *Control of Transcription Termination,* Routledge, New York, 1993.

Davidson. J. N. : *Biochemistry of Nucleic Acids,* Cox and Wyman, London, 1965.

Demidov, V. V. : *PNA and LNA throw Light on DNA, Trends,* Oxford, New York, 2003.

DuPraw, E. J. : *Cell and Molecular Biology,* Academic Press, New York, 1969.

Ebby, S. R. : *Non Coding RNA Genes and Modern RNA,* Oxford, New York, 2001

Echols, H. : *Fidelity Mechanisms in DNA Replication,* Oxford, New York, 1991.

Ellis, R. J. : *Chloroplast Protein Synthesis,* Oxford, New York, 1975.

———— : *Molecular Chaperones,* Oxford, New York, 1991.

———— : *The Biogenesis of Chloroplasts Protein,* Oxford, New York, 1974.

Fischer, L.M. : *DNA Unwinding in Transcription and Recombination,* Oxford, New York, 1982.

Friedberg, E. C. : *Relationships Between DNA Repair and Transcription,* Oxford, New York, 1996.

Gesteland, R. F. and J. F. Atkins : *The RNA World,* Oxford, New York, 1993.

Gilbert, W. and B. Muller : *Isolation of the Lac Repressor,* Oxford, New York, 1966.

Gilham, N. W. : *Genetic Analysis of the Chloroplast and Mitochondrial Genomes,* Oxford, New York, 1974.

Gilmour, R. S. and J. Paul : *Tissue Specific Transcription of the Globin Gene in Isolated Chromatin,* Oxford, New York, 1973.

Glick, B. and G. Schatz : *Import of Proteins into Mitochondria,* Oxford, New York, 1991.

Gray, M. W. : *The Evolutionary Origin of Organelles,* Oxford, New York, 1989.

Grunstein, M. : *Histones as Regulators of Genes,* Oxford, New York, 1992.

Gupta, P. K. : *Cell and Molecular Biology*, Rastogi Publications, Meerut, 2000.

Hadjiolov, A. A. : *The Nucleolus and Ribosome Biogenesis*, Oxford, New York, 1985.

Hoagland, M. G. : *The Present Status of the Adaptor Hypothesis*, MacMillan, London, 1959.

Hoeijmakers, J. H. : *Nucleotide Excision Repair from Yeast to Mammals*, MacMillan, London, 2001.

Holley, R. W. : *Structure of a Ribonucleic Acid*, McGraw Hill, London, 2002.

Ingram, V. : *The Biosynthesis of Macromolecules*, Benjamin, New York, 2001.

Jacob, F. : *Genetic Regulatory Mechanisms in the Synthesis of Proteins*, Oxford, New York, 1961.

Jukes, T. H. : *Mitochondrial Codes and Evolution*, Oxford, New York, 2000.

Kaguni, L. S. : *Eukaryotic DNA Polymerase-primase : Structure, Mechanism and Function*, MacMillan, London, 1999.

Kaziro, Y: *Structure and Function of Signal Transducing GTP Binding Proteins*, Blackwell, London, 2001.

Kendrew, J. C. : *The Three-dimensional Structure of a Protein Molecule*, Sinauer, New York, 1991

Kornberg, A. and Baker, T.A. : *DNA Replication*, Freeman, New York, 2002.

Kramer, A. :*The Structure and Function of Proteins Involved in Mammalian Pre-mRNA Splicing*, Grandada, New York, 2001.

Kurland, C. G. : *Structure and Function of the Bacterial Ribosomes*, Oxford, New York, 2000.

Lake, J. A. : *The Ribosome*, Oxford, New York, 1981.

Lehninger, A. L. : *The Mitochondrion*, Academic Press, New York, 2001

Lewin, Benjamin : *Genes*, Oxford University Press, New York, 2000.

Maxwell, E. S. and Fournier, M.J. : *The Small Nucleolar RNAs*, MacMillan, London, 1995.

McKnight, S. L. : *Molecular Zippers in Gene Regulation*, Princeton, New York, 2002.

Merrick, W. C. : *Overview : Mechanism of Translation Initiation in Eukaryotes*, Longman, London, 2001.

Moore, P. B. : *Molecular Mimicry in Protein Synthesis*, Cambridge, London, 2002.

Noller, H. F. : *Ribosomol RNA and Translation*, Oxford, New York, 1999.

Olson, R. E. : *Enzymatic Synthesis of DNA*, John Wiley, New York, 2001.

Owen, L. :*DNA Synthesis*, Freeeman, San Francisco, 2001.

Pabo, C. O. and R.T. Sauer : *Transcription Factors : Structural Families and Principles of DNA Recognition*, Printice Hall, London, 1993.

Pardee, A. B. : *Molecular Basis of Biological Regulation: Origins of Feedback Inhibition and Allostery*, Arnold, London, 2003.

Pardue, M. L. and J. G. Gall : *Molecular Hybridization of Radioactive DNA to the DNA of Cytological Preparations*, Dekker, New York, 2001.

Paszkowski, J. : *DNA and Histone Methylation in Plants*, MacMillan, London, 2002.

Razin, A. and A. D. Riggs : *DNA Methylation and Gene Function*, Longman, New York, 2002.

Rich, A. and S. H. Kim : *The Three Dimensional Structure of Transfer RNA*, Oxford, New York, 1999.

Robertson, M. : *RNA Processing : the Post-RNA World*, Oxford, New York, 1999.

Rosenberg, M. and D. Court : *Regulatory Sequences Involved in the Promotion and Termination of RNA Transcription,* Cambridge, New York, 2003.

Sager, R. : *Cytoplasmic Genes and Organelles,* Academic Press, New York, 2003.

Schmid, M. B. : *Structure and Function of the Bacterial Chromosome,* Granada, London, 1990.

Spector, D. L. : *The Dynamics of Chromosome Organization and Gene Regulation,* Oxford, New York, 2005.

Stryer, L. : *Biochemistry,* W.H. Freeman & Co., New York, 2002.

Temin, H. M. : *RNA Directed DNA Synthesis,* MacGraw-Hill, New York, 2002.

Turner, B. M. : *Histone Acetylation and Control of Gene Expression,* MacMillan, London, 2002.

Ulrich, B. : *Mechanism and Regulation of Eukaryotic Protein Synthesis,* Oxford, New York, 2002.

Uptain, S. M. : *Basic Mechanisms of Transcript Elongation and its Regulation,* MacMillan, London, 1997.

Vernadsky, V. I. : *Organellar Heredity,* Revan Press, New York, 1978.

Voinnet, O. : *RNA Silencing : Small RNAs as Ubiquitous Regulators of Gene Expression,* MacMillan, London, 2002.

Wallace, D. C. : *Diseases of the Mitochondrial DNA,* Blackwell, New York, 2002.

Wang, T. S. F. : *Eukaryotic DNA Polymerases,* MacMillan, London, 2002.

Watson, J. D. : *Molecular Biology of the Gene,* Oxford, New York, 2002.

———— : *Molecular Structure of Nucleic Acids : a Structure for Deoxyribose Nucleic Acid,* Oxford, New York, 2001.

Wilkins, M. H. F. : *Physical Studies of the Molecular Structure of Deoxyribose Nucleic Acid and Nucleoprotein,* MacMillan, London, 2002.

Williams, R. C. : *Reconstitution of Active Tobacco Mosaic Virus from its Inactive Protein and Nucleic Acid Components,* Oxford, New York, 1955.

Williamson, B. : *Split Gene Transcription,* Oxford, New York, 2001.

Woese, C. R. : *The Genetic Code : The Molecular Basis for Genetic Expression,* Harper & Row, New York, 2003.

Wuthrich, K. : *Protein Structure Determination in Solution by Nuclear Magnetic Resonance Spectroscopy,* Oxford, New York, 2001.

Yamamoto, T. : *A Bipartite DNA Binding Domain Composed of Direct Repeats in the TATA Box Binding Factor TFHD,* Oxford, New York, 2001.

Yamashita, S. : *Transcription Factor TFHB Sites Important for Interaction with Promoter-bound TFIID,* Springer, London, 2000.

Yams, M. : *Translational Efficiency of Transfer RNAs : Uses of an Extended Anticodon,* MacMillan, London, 2003.

Yanofsky, C. : *The Tryptophan Synthetase System,* Oxford, New York, 2001.

Zawel L., and D. Reinbcrg : *Common Themes in Assembly and Function of Eukaryotic Transcription Complexes,* MacMillan, London, 2002.

Zuckerman, B. : *Initiation of mRNA Translation in Prokaryotes,* Oxford, New York, 1990.

Zurick, D. : *Origin and Evolution of Mitochondrial DNA,* Oxford, New York, 1989.